TEXANS ON THE BRINK

INTEGRATIVE NATURAL HISTORY SERIES

Sponsored by Texas Research Institute for Environmental Studies, Sam Houston State University

WILLIAM I. LUTTERSCHMIDT

AND BRIAN R. CHAPMAN, General Editors

Texans

on the Brink

THREATENED AND ENDANGERED ANIMALS

Edited by BRIAN R. CHAPMAN *and* WILLIAM I. LUTTERSCHMIDT

Foreword by JOHN H. RAPPOLE

TEXAS A&M UNIVERSITY PRESS College Station

Library of Congress Cataloging-in-Publication Data

Names: Chapman, Brian R., editor. | Lutterschmidt,
  William I., editor.
Title: Texans on the brink: threatened and endangered
  animals / edited by Brian R. Chapman and William I.
  Lutterschmidt; foreword by John H. Rappole.
Description: College Station: Texas A&M University Press,
  [2019] | Series: Integrative natural history series | Includes
  bibliographical references and index. |
Identifiers: LCCN 2018036559 (print) | LCCN 2018037759
  (ebook) | ISBN 9781623497323 (ebook) | ISBN 9781623497316 |
  ISBN 9781623497316 (printed case: alk. paper)
Subjects: LCSH: Endangered species—Texas. | Animal
  diversity conservation—Texas.
Classification: LCC QH76.5.T4 (ebook) | LCC QH76.5.T4 T477
  2019 (print) | DDC 591.6809764—dc23
LC record available at https://lccn.loc.gov/2018036559

With appreciation for the biologists,

landowners, and lawmakers dedicated to

understanding, managing, and protecting

the biological diversity of the state and

nation.

To keep every cog and wheel is

the first precaution of intelligent

tinkering.

— ALDO LEOPOLD

CONTENTS

FOREWORD

When you turn the management [of federal lands] over to the tree-huggers, the bird and bunny lovers and the rock lickers, you turn your heritage over.
— MIKE NOEL, Utah state representative, *New York Times*, 24 August 2017

The above quote shows clearly the timeliness of the superb summaries of the endangered species of Texas presented in this volume. As explained by coauthors Brian Chapman and William Lutterschmidt in chapter 3, endangered species are on the front lines in the war to preserve America's spectacular natural legacy. In fact, since enactment of the Endangered Species Act of 1973, they have been the shock troops, a role in which they have been extraordinarily effective. One need think only of the Snail Darter, Spotted Owl, Brown Pelican, Peregrine Falcon, Bald Eagle, and Texas' own Golden-cheeked Warbler to recall tough battles fought and won by these fragile warriors. Times change, though, and victory that seemed so recently within our grasp appears further away now than at any time in the past 44 years. Mr. Noel, who is currently (August 2017) under consideration to lead the Bureau of Land Management, presents a message to those who would try to protect this country's environmental birthright that could hardly be more stark—namely, that anyone who does not agree with unbridled exploitation of the West's natural resources is an effete outsider whose mission is subversion of the pioneer spirit. Many Texas politicians evidently agree, as demonstrated by their 2011 transfer of responsibility for the state's endangered species program from the Texas Parks and Wildlife Department to the Comptroller of Public Accounts.

The ironies in this naked attempt to relegate the champions of environmental legal combat to insignificance are overwhelming, not the least of which is that Texas has more to lose in this conflict than any other state in the country. Because Texas is special. Of course Texans take pride in their state's uniqueness, but most have little understanding of just how exceptional it is from a natural history perspective. The contents of this book should make this preeminence perfectly clear, but it warrants special emphasis. People often mention size as the most critical factor in what makes the state special in terms of its beauty and diversity. But size alone is insignificant. What makes Texas matchless are its forests, deserts, rivers, mountains, and seas straddling the temperate and subtropical realms of the continent. These elements make Texas the richest natural area in North America north of the tropics; richer than all of the other states and Canada combined if one excludes California; indeed, it is far richer in its plant and animal life than the entire continent of Europe. To further emphasize the importance of the quality and diversity of habitats rather than the quantity, the tiny Texas Coastal Bend region, which contains only 11 of the state's 254 counties, an area smaller than the state of Massachusetts, contains about half of the state's plant and animal diversity.

A further irony is the implication that any attempts to enjoy and preserve this natural inheritance are somehow at odds with the state's cultural heritage. Some of the greatest names in Texas culture have often led the way in efforts to preserve the state's beauty and diversity. For example, the King Ranch, a Texas cultural icon if ever there was one, has been an effective and outspoken leader in preservation of the state's flora and fauna. For instance, the ranch established the Caesar Kleberg Wildlife Research Institute mainly for the study and preservation of biodiversity, and subsequently, through the leadership efforts of Sam Beasom and Fred Bryant, enlisted the aid of a number of Texas cattlemen in that effort. The ranch also led the fight to prevent channelization of the vast Laguna Madre, a wetland larger and more diverse than the Everglades. Similarly, cattleman Rob Welder established the Rob and Bessie Welder Wildlife Foundation for the advanced education of wildlife and environmental professionals, an organization led during its first two decades by Dr. Clarence Cottam, who left his post as dean of Utah's Brigham Young University to lead Welder to prominence as a preeminent Texas supporter of training for wildlife professionals. These exemplary models of Texas culture leading the effort to preserve its natural heritage represent the large number of outstanding cultural leaders in Texas who have also cared deeply about its natural legacy.

The vast majority of Texans want their natural heritage preserved. This book, authored by 41 leading experts who have devoted their professional lives to its study, provides excellent information on why it's important. Ultimately, however, it will be up to the state's citizens to grasp the nettle. They must hold their elected officials accountable if anything is to be done to keep the 100 taxa treated here from joining the Ivory-billed Woodpecker, Carolina Parakeet, and Passenger Pigeon in oblivion.

John H. Rappole
Research Scientist Emeritus
Smithsonian Conservation Biology Institute
Front Royal, Virginia

EDITORS' PREFACE

On June 24, 1519, Alonso Álvarez de Pineda sailed into an uncharted bay teeming with life. Schools of huge Atlantic Tarpon plied the waters, Caribbean Monk Seals frolicked in the pass to the Gulf, and West Indian Manatees grazed on lush Laguna Madre seagrasses. After a brief circumnavigation of the anchorage now known as Corpus Christi Bay, de Pineda sailed on to map more of the coastline along the western Gulf of Mexico. If he had disembarked, he might have caught a glimpse of an Attwater's Prairie-Chicken, an elusive Ocelot, or perhaps a skulking Jaguarundi in the thickets near the coast. If he had lingered in the area until fall, he would have witnessed thousands of Eskimo Curlews feeding in the shallows of bays and estuaries, joined by immense flocks of other shorebirds and waterfowl. Farther inland, he surely would have been enthralled by the herds of massive American Bison on the plains and savannas, impressive Ivory-billed Woodpeckers in steeple-high Longleaf Pine forests and Baldcypress swamps, and trees near the Red River laden with thousands of roosting Passenger Pigeons.

During the five centuries since de Pineda's explorations, Jaguarundis and West Indian Manatees have been extirpated from Texas, and the few remaining Ocelots and Attwater's Prairie-Chickens, both listed as endangered, now occupy protected refuges and zoos devoted to their survival. As few as 25 American Bison, which once numbered in the tens of millions, survived a period of widespread slaughter, and their descendants are now confined to wildlife refuges, parks, and private ranches. The Caribbean Monk Seal, Ivory-billed Woodpecker, and Passenger Pigeon are extinct, and the Eskimo Curlew likely is as well, having not been seen since 1986. Although many other species experienced decline or extinction after de Pineda's departure, the beauty and remaining biodiversity of the state continue to attract tourists, naturalists, and hunters.

While no one can bring back the extinct species that contributed to the biodiversity of the Lone Star State, we hope that this book will provide the reader with current and detailed understanding of each of the 88 animal species designated as threatened, endangered, or conditional by the US Fish and Wildlife Service or the National Marine Fisheries Service.

This is the second book to describe the threatened and endangered animals of Texas. The first, published in 1996 by the Texas Parks and Wildlife Press, was written by Linda Campbell. That landmark volume was the first to describe the portion of the state's rich natural heritage that was vulnerable to extinction. It described the life history and habitat of each species listed by the US Fish and Wildlife Service as threatened or endangered and offered useful information to interested Texas citizens, especially educators, landowners, and land managers. All Texans, especially those of us laboring daily to conserve vulnerable species, owe a debt of gratitude to Ms. Campbell for her monumental effort to enlighten and engage the public.

During the 20-plus years since the afore-mentioned volume was published, the state's list of threatened and endangered species has changed; many species have been added to the list and a few have been deleted (because of recovery or extinction). Through their research, biologists have also learned much more about the biological and ecological requirements of some listed species and have altered their management efforts. State, federal, and nongovernmental organizations have set aside many lands throughout the state to safeguard the habitats critical to certain species. With these developments in mind, we decided it was time to produce a new volume.

Although our ultimate goal—providing current, accurate, and useful information—mimics that of the first volume, our approach differs. To ensure that the information about each species described in this book is as accurate as possible, we enlisted the aid of professional biologists to write the species accounts. Each of these authors is an expert intimately familiar with the species or group of species described, and each generously contributed his or her expertise. At our request, the authors included a portrayal of the environmental concerns that must be addressed to ensure the continued existence of the species in their accounts. The absolute correlation between the protection of habitat and the conservation of wildlife is obvious in each species account.

This volume also differs by offering the reader an overview of biodiversity values. The long and complicated process by which an animal is added to the federal threatened and endangered list is described in chapter 4. After a species is proposed for listing, biologists collect and evaluate the available biological and ecological data, conduct field studies, consult experts, receive public comments, and reevaluate the accumulated data. We hope that our description of the required processes and the time and care involved will help expand public understanding that the listing of a species is neither frivolous nor easy. In fact, as we write this, hundreds of species proposed for listing are waiting for their status to be assessed. Many will eventually be added to the list of threatened and endangered species, but some may go extinct in the interim. There simply are not enough biologists or funds available in the responsible federal agencies to assess the vulnerability of all the species edging nearer to the eternal abyss of extinction with their feet on the proverbial banana peel.

We have no doubt that this volume will have to be updated in the future. Before that happens, it is our hope that many more species will be removed from the threatened and endangered list because their habitats have been restored and their populations have recovered than because they have disappeared forever. We therefore dedicate this book to the many biologists and citizens who devote their knowledge, skills, and financial means to the preservation of biodiversity.

Brian R. Chapman and
William I. Lutterschmidt
Authors and editors
Texas Research Institute for
Environmental Studies
Sam Houston State University,
Huntsville, Texas

ACKNOWLEDGMENTS

This compilation could not have been finalized without generous contributions of information, advice, and expertise by the authors of the various species accounts and by colleagues who eased our editorial tasks. For their advice and assistance, our thanks are extended to Kathy Allen, Eric G. Bolen, Julia C. Buck, Sandra S. Chapman, Jerry L. Cook, Craig Farquhar, Neil Ford, Ron R. George, William B. Godwin, David Hoffpauir, Ashley Morgan, Paige A. Najvar, Cliff Shackelford, and Autumn Smith-Herron.

This book is the third contribution to the Integrative Natural History Series sponsored by Sam Houston State University (SHSU) and published by Texas A&M University Press. The Texas Research Institute for Environmental Studies and the SHSU Office of Research and Sponsored Programs (ORSP) provided financial resources that supported the development and production of this book. We are particularly grateful for the encouragement and assistance offered by Richard Eglsaer, provost and vice president for academic affairs, and Jerry L. Cook, associate vice president of ORSP. Without their support and guidance, neither the book series nor the book itself would have been possible.

Individuals and agencies that provided photographs are acknowledged in the appropriate figure captions, but this seems hardly enough thanks for their generosity and willingness to share the fruits of their labor. We are indebted to those who offered their photographs for public use through Creative Commons—without access to those images, the pages of this book would be barren indeed. Photography of tiny cave-adapted invertebrates requires perseverance, special equipment, and singular talent, and we are grateful to Dr. Jean Krejca of Zara Environmental for the use of the many images that populate chapter 6.

Constructive comments by external reviewers provided insights and information that we appreciate. We are grateful for the observations, corrections, and ideas provided by Janice Bezanson, Meredith Longoria, and Randy Simpson. Sandra S. Chapman painstakingly proofread many of the chapters, and we appreciate her invaluable suggestions. Shannon Davies, Stacy Eisenstark, and Katie Duelm provided invaluable editorial supervision of the book at Texas A&M University Press. The final text of this book benefited from the masterful copyediting of Laurel Anderton, and we are grateful for her assistance. Thanks to Sherry Smith for the index. Although many contributed to this book, we alone accept all responsibility for any errors that escaped our notice.

We are especially indebted to John H. Rappole, who graciously wrote the foreword for this compilation. As a research scientist with the Conservation and Research Center of the Smithsonian Institution's National Zoological Park, John has written many papers and books that have greatly expanded the scientific understanding of avian migration ecology and the significance of maintaining suitable habitats in Mexico and the United States. We are honored by his willingness to support this book.

Finally, we cannot forget to acknowledge the contributions of our wives, Sandy and Catherine, who persevered while we worked on this project. Their encouragement, support, and understanding nourished our creativity and kept us on track.

Brian R. Chapman
Senior Research Scientist
Texas Research Institute for
Environmental Studies
William I. Lutterschmidt
Executive Director of Research Laboratories,
and Professor of Biological Sciences

PART ONE

Conservation of Biodiversity

1 WHY SAVE BIODIVERSITY?

BRIAN R. CHAPMAN & WILLIAM I. LUTTERSCHMIDT

We should preserve every scrap of biodiversity as priceless as we learn to use it and come to understand what it means to humanity. — E. O. WILSON

Several decades ago, I (BRC) appeared on a locally produced television talk show to describe the interesting plants and animals of South Texas. When the discussion turned to at-risk species, the host asked me to name a few rare animals that might someday be threatened or endangered. Upon mention of the Western Massasauga in the list of vulnerable species, the host immediately interrupted to ask, "What is a Western Massasauga?" When I said it was a small rattlesnake, the host immediately responded, "What good are rattlesnakes?" This question, in various forms, is frequently asked of biologists who work with rare or obscure species, especially those that represent a potential danger to humans. Many people who do not have a close association with nature ask a similar question: "Why should I be concerned about something that I have never seen?"

The potential loss of a large or charismatic species like the Bald Eagle, American Bison (fig. 1.1), or Whooping Crane generates passionate public concern, partially because these animals are widely recognized, and we tend to place more value on creatures that are more familiar. Clearly, the obvious disappearance of a well-known or cherished species also signals that something is drastically wrong. Unfortunately, the same concern is rarely engendered for small creatures that remain out of public view. It is extremely difficult to generate public anxiety about the potential extinction of a cave spider, a tiny salamander, or a guppy-like fish. Even more problematic is the task of garnering interest in saving a small, venomous animal like a massasauga.

Nature is composed of many assemblages of plants and animals that coexist and interact to form recognizable associations called biological communities (biotic communities, for short). Although many citizens know little about most of the individual species that make up biotic

Figure 1.1. Unregulated hunting almost eliminated the American Bison from North America. The species is now recognized as the National Mammal of the United States. Photograph by Sandra S. Chapman.

communities, most recognize the associations by their general names—who among us cannot visualize a prairie, a forest, or a desert? The species interacting within each of these biotic communities represent the living components of an ecosystem. The variety of species within an ecosystem is often termed species diversity, biological diversity, or biodiversity (a shortened form). Within an ecosystem, the plants and animals of a biotic community interact with and depend on the physical components of the environment—the soil, nutrients, water, and air. Each component plays a role in maintaining the complexity and resilience of a prairie, forest, or other type of biotic community. Envision a large, dew-encrusted spiderweb glistening in the morning sun across a forest opening (fig. 1.2). Each intersection of gossamers forming the web can be imagined to represent a species in an ecosystem. As long as the intersections all remain intact, the spiderweb remains strong and its ability to capture a moth or fly is ensured. As holes are poked in the web, the junctures are compromised, weakening the entire web until it eventually collapses. In a corresponding manner, when a species disappears from the web of an ecosystem, the biotic

Figure 1.2. The jewel-like dew droplets encrusting this spiderweb enhance the junctions reinforcing the structure. The web's inter-sections are often likened to species in an ecosystem. Photograph courtesy of Creative Commons.

Figure 1.3. The removal of some predators, such as this sea star, can result in a trophic cascade. Photograph courtesy of Creative Commons.

community becomes less sustainable for other species, including humans.

So, what good are rattlesnakes? As carnivores, rattlesnakes, including massasaugas, reduce populations of small mammals, especially rodents. The most obvious consequence of removing rattlesnakes from any ecosystem is a population explosion of their prey species. Although they prowl slowly at night, rattlesnakes are efficient predators that enter spaces such as burrows and crevices where foxes or Coyotes cannot go. Rats and mice can carry fleas, ticks, and disease organisms easily transmitted to humans, especially when rodent populations are high. Infestations of rodents can also damage vegetation, including food crops, as well as stored supplies, electric wiring, and structures. Rattlesnakes, especially small ones, are not the top predator in a food chain—they also serve as prey for birds like hawks

or herons and mammals like skunks or raccoons. When a rattlesnake dies, the nutrients within its body return to the soil through decomposition aided by bacteria, fungi, and worms. Throughout its life and even in death, a Western Massasauga is a positive actor in the biotic community.

Nature has provided numerous examples of how the removal of a single species from an ecosystem can produce noticeable—and sometimes unexpected—results. The interrelationships among all species in most ecosystems have not been evaluated, so it often comes as a surprise when an obscure species represented by a relatively small population plays a significant role in maintaining balance among species. For example, when a diminutive sea star (fig. 1.3), a key predator of mussels, was removed from a tidal plain in Makah Bay, Washington, the mussel population exploded, eliminating many other species in a chain reaction called a trophic cascade. Because of its size and relatively low numbers, the importance of the petite predator in limiting population explosions of mussels and maintaining ecosystem stability was overlooked.

The role of the tiny sea star in this ecosystem was analogous to that of a keystone in a rock archway. Removal of a keystone braced against both sides of an arch will result in collapse of the entire structure. Keystone species play a similar role in an ecosystem—their presence and activities profoundly influence the structure of the community in which they live. Reduction or elimination of such species often results in dramatic and unpredictable ecosystem disruption. As biologists expose the roles of species in additional ecosystems, the importance of more keystone species will be discovered. Many of these initially appear inconsequential and are easily dismissed by a casual observer.

Unfortunately, the frequency of extinctions is increasing at an alarming rate, and many species may disappear before their role or importance in an ecosystem becomes known. Scientists fear that we are currently experiencing a mass extinction event—a phenomenon in which a large portion of the world's species die off. Populations of animals all over the globe are declining so rapidly that researchers describe the process as "biological annihilation." Mass extinctions have happened on five previous occasions. The most recent occurred about 65 million years ago when a huge meteor crashed into the Earth and caused the extinction of the dinosaurs. The current event began about 200 years ago as the expanding human population began to alter the environment in a significant manner. During the past two centuries, humans have built great cities on top of animal habitats, plundered the ocean's fisheries, dammed rivers, plowed under grasslands, harvested entire forests, expanded deserts, released toxic pollutants, and altered the atmospheric concentration of greenhouse gases. Consequently, more than 500 species, subspecies, and varieties of our nation's plants and animals have become extinct since the Pilgrims set foot on Plymouth Rock in 1620. The present extinction rate may be up to 100,000 times higher than the rate that existed over the millennium prior to 1800, and nearly 50 percent of the animal species that were present 200 years ago have already disappeared.

The importance of biodiversity preservation seems difficult for many people to grasp. Nonetheless, support for conservation of species and biotas, the intrinsic value of nature, and the significance of species to ecosystems cannot be stressed enough. We easily observe and experience the impacts of water and air pollution, but we often overlook the value of ecosystem complexity to our everyday life. People seem to be losing their awareness of nature and the ecosystem services that "Mother Nature" provides. How many spider species live in your flowerbed and protect your garden from pests? Would you know if one of these species became extinct? Animal populations often diminish in subtle increments, and some species disappear without our even knowing they were once present. What follows are just a few of the many reasons humanity must help preserve as many species and ecosystems as possible.

Human dependence on ecosystem functions was demonstrated recently when populations of the European Honey Bee began to decline worldwide (fig. 1.4). From 2006 to 2013, more than 10 million beehives were lost to a condition known as "Colony Collapse Disorder." When a honey bee colony collapses, the workers disappear, leaving

Figure 1.4. Pollination of flowers and many food crops is essential to ecosystem stability and human food production. Note the pollen sacs on this Honey Bee's legs. Photograph by Bob Peterson, Creative Commons.

an insufficient workforce to feed and care for the developing larvae. The disorder causes significant economic losses because many agricultural crops, including about one-third of US crops, are pollinated by bees. Grain crops, which are wind pollinated, are not affected by the disappearance of honey bees, but many fruits, berries, beans, melons, nuts, and citrus crops depend on insect pollination. Billions of dollars were lost in agricultural markets worldwide because of the collapse of honey bee populations.

Many arguments for conserving species emphasize the potential value of "undiscovered riches." An especially tangible benefit of biodiversity is its contributions to medicine and human health. Natural chemicals taken directly from or modeled on those found in plants and animals are used to prepare approximately 50 percent of our current prescription medicines. Every living species contains a trove of unique genetic material representing the product of eons of evolution. Some of the most unlikely organisms produce molecules that are useful in fighting human diseases. For example, derivatives of the brevetoxins produced by Red Tide organisms, which are common in waters of the Texas Gulf Coast, may eventually treat cystic fibrosis, a severe lung disorder. More recently, Harvard University researchers discovered that derivatives of the

Figure 1.5. Millions of Brazilian Free-tailed Bats exit Central Texas caves nightly to feed on aerial insects, including mosquitos and many agricultural pests. Photograph by dizfunkshinal, Creative Commons.

feed on and decompose animal wastes, converting them into soil-enhancing fertilizers. Brazilian Free-tailed Bats exit caves in Central Texas nightly during the summer months to feed on airborne agricultural pests, including moths, lacewings, armyworms, and bollworms (fig. 1.5). Free services rendered by the bats significantly reduce populations of insect pests, increase agricultural productivity, diminish the need for pesticides, and save farmers millions of dollars annually. Freshwater mussels serve as nature's water filters, constantly removing contaminants and particulate matter from the state's rivers and streams.

Finally, the beauty of our planet is a by-product of biodiversity. Many people travel widely about the globe to enjoy natural surroundings and wildlife. Imagine how much less exciting a trip to Africa would be if African Elephants, African Lions, Cheetahs, or Reticulated Giraffes were no longer part of the savanna ecosystem. The same would be true for a trip to Yellowstone National Park if American Bison, Elk, or Gray Wolves no longer existed. The aesthetic and economic values of biodiversity, however, are not confined to the nation's large national parks. The Texas Parks and Wildlife Department estimates that birders who participate in "the nation's fastest growing outdoor recreation" pump about $400 million annually into the state's economy. Biodiversity provides some of the most wonderful aspects of our existence and adds immensely to the quality of our lives.

From 1911 until 1986, British coal miners carried caged Canaries into the mine with them. More than just pets, the Canaries served an important function—sentinels. These sensitive birds succumbed to carbon monoxide and other toxic gases in the mine well before the vapors reached levels affecting human survival. When a Canary died, it was time to evacuate the mine. Perhaps the rapid disappearance of animal species provides a similar warning. *Every species is worth saving.*

slimy yellowish mucus produced by the Dusky Arion, a terrestrial slug regarded as a garden pest, can be used as a medical superglue to seal wounds. The glue, which adheres to wet surfaces, has potential lifesaving uses in heart surgery and cases of internal bleeding. To date, scientists have investigated only a small fraction of the world's species to discover their secrets.

Biodiversity contributes enormous economic value to the citizens of Texas. Plant and animal products are the primary sources of our food and energy. Certain plant species remove heavy metals or other contaminants from the soil, water, or air, and some species enhance and maintain fertile soils by their biological processes. Most green plants reduce greenhouse gases, the major source of global warming, by sequestering atmospheric carbon. Numerous species of insects and worms

2 CHARACTERISTICS OF VULNERABLE SPECIES

BRIAN R. CHAPMAN & WILLIAM I. LUTTERSCHMIDT

The whale is endangered, while the ant continues to do just fine. — BILL VAUGHN

We are all familiar with extinction. Visitors to many natural history museums are greeted by displays of dinosaurs, mammoths, and other fossilized remains of animals that once roamed widely across North America (fig. 2.1). These impressive species flourished until an ancient occurrence altered the environment at a rate that outpaced their ability to adapt. Prehistoric natural events such as volcanic eruptions, asteroid impacts, or competition with more successful species caused the extinction of species at an average rate of about five per year. Around AD 1600, however, humans became the primary cause of species extinctions, and their impacts accelerated the rate of species loss. Advances in agriculture, industrial development, urban expansion, and the rapid growth of human populations generated new environmental challenges for the world's biota. The current rate of species die-offs represents the worst epidemic of extinctions since an asteroid eradicated dinosaurs and countless other animals 65 million years ago.

Many state, federal, and international agencies now monitor animal populations within their jurisdiction to develop comprehensive lists of species nearing extinction. For example, the International Union for Conservation of Nature (IUCN), located in Morges, Switzerland, obtains information on endangered species from over 1,700 member agencies representing more than 130 countries. It publishes the description and current status of each species in the *IUCN Red List of Threatened Species*. Analysis of the information in the list reveals that four main factors are responsible for most current losses.

NATURAL CAUSES. Millions of species that evolved over the millennia eventually ceased to exist as a result of a natural catastrophe, such as a volcanic eruption or rapid environmental change. Others disappeared as a natural outcome of the evolutionary process in which overspecialized

Figure 2.1. *Tyrannosaurus rex*, a dinosaur familiar to most schoolchildren, roamed across the Texas landscape until an asteroid impact eliminated all large reptiles about 65 million years ago. Photograph of the holotype, displayed at the Carnegie Museum of Natural History, Philadelphia, by Scott Robert Anselmo, Creative Commons.

species were replaced by a more generalized and efficient competitor. Through the ages, millions of species have flourished for a time before winking out because of some natural cause. Unfortunately, only those species with hardened features leave lasting evidence of their existence through fossils. Based on fossil evidence, the average "life span" for a bird species is about 2 million years, and for a mammal species only 600,000 years. Natural causes still eliminate some species, but they do not contribute significantly to the current rate of extinctions.

UNREGULATED HUNTING. The annihilation of Passenger Pigeons, once considered the most abundant avian species in North America, may represent the most well-known extinction resulting from unregulated hunting. The disappearance of the colorful Carolina Parakeet, the eradication of

Figure 2.2. The Black-footed Ferret, an inhabitant of Black-tailed Prairie Dog colonies, no longer occurs in Texas and remains vulnerable to extinction elsewhere in western North America. Photograph by Ryan Hagerty, US Fish and Wildlife Service.

the Jaguar from West Texas, and the near extinction of the American Bison also serve as testimonials to the decimation that can result from unregulated hunting. The elimination of a key species may also affect populations of nontarget species. For example, Black-footed Ferrets, once a common occupant of Black-tailed Prairie Dog colonies, were eliminated from Texas and nearly became extinct elsewhere (fig. 2.2). Their populations declined as prairie dogs, which once numbered in the hundreds of thousands, were eradicated from large portions of their former range by gunshot, poisoning, and gassing.

INTRODUCTION OF NONNATIVE ORGANISMS.
Species composition in a biotic community is achieved through generations of interactions involving interspecific competition—contests between two or more species to acquire the same limited resource. Over time, either one or more of the species dies off or each species modifies its use of the resource, be it food, space, or nesting sites.

When a nonnative species (an "exotic" species) is introduced by human actions, the delicate balance of an ecosystem is often disrupted. Exotic species usually enter an environment devoid of natural controls (predators and parasites) that keep populations in check. In the absence of limiting factors, the introduced species reproduces more successfully than native species, and its population expands rapidly. The exotic organisms often enter new areas and become known as "invasive species." Invariably, some native species

are displaced or eliminated. European Starlings, for example, displace cavity-nesting birds such as the Eastern Bluebird by occupying and defending the cavity first, or by tossing out the eggs or nestlings of a nesting bluebird. Exotic predators are especially devastating. The Brown Anole, an invasive species native to Cuba and the Bahamas, was introduced via the pet trade into Florida and other southern states, where it established populations. The lizard was subsequently shipped to Texas in plants and established populations in South Texas, where it preys on young Green Anoles, a native species. In areas where Brown Anoles are abundant, populations of Green Anoles are declining or completely absent (fig. 2.3). One of the tiniest invasive species, the Red Imported Fire Ant, has had a devastating impact on a variety of organisms ranging from insects to mammals, and its range continues to expand across the southern half of the United States.

Nonnative plants can be just as disruptive as exotic animals. After Turnip Weed (a.k.a. Bastard Cabbage) was introduced from Europe many years ago, it began to dominate many Texas roadsides and fields. It grows larger and faster than most native wildflowers and gains an extra advantage by germinating in the fall and overwintering as a basal rosette. Given this head start over spring-germinating wildflowers, it shades out bluebonnets, Indian Blankets, and other colorful native species. In some areas, Bastard Cabbage has transformed the roadside coat of many colors into monotonous stretches of yellow at the expense of the native species. Other nonnative plants have degraded rangelands, eliminating nesting habitats for Northern Bobwhites and other grassland birds.

HABITAT ALTERATION. Without question, the modification or complete destruction of wild land resulting from human activity is the most significant cause of species declines and extinctions. Clear-cutting in the pine savannas and swamps of East Texas eliminated the habitat for Ivory-billed Woodpeckers and limited survival chances for the Red-cockaded Woodpecker and Louisiana Pinesnake. Sand mining along rivers introduced fine silts that reduced populations of Golden Orbs and other freshwater mussels.

Figure 2.3. The native Green Anole (left) is being eliminated by the Brown Anole (right), a nonnative invasive species imported from the West Indies. Photographs courtesy of R. Scott Blenis (left) and Hans Hillewaert (right), Creative Commons.

Clearing of acreage for homesteads and farms allowed Coyotes to expand their range eastward and hybridize with Red Wolves, which nearly eliminated Red Wolves as a distinct species. Urban expansion, reservoir construction, draining of underground aquifers for agricultural and urban uses, highway construction, and chemical pollution are just a few of the many other examples of human developments that contribute to species declines.

What Predisposes a Species to Vulnerability?

A few species possess traits that allow them to persist or even thrive in the face of significant pressures on their population. After over a century of campaigns to eliminate them by hunting and poisoning, Coyotes offer a prime example of such adaptability (fig. 2.4). Their population remains sound in their original range, and the species has expanded as far east as the Atlantic Coast and northeastern Canada. Many species, however, possess one or more characteristics that predispose them to population declines. Even the slightest

changes in climate, habitat, or composition of their biological community may be enough to push these species toward endangerment. These characteristics include the following:

- Species with restricted distributions: The Austin Blind Salamander inhabits only three spring outlets in Austin's Zilker Park and the underground aquifer that feeds the springs. The San Marcos Salamander is restricted to the spring outlets and surrounding submergent vegetation at the bottom of Spring Lake in San Marcos. Drying of these springs because of groundwater depletion or prolonged drought would eliminate the habitat for these species. Similarly, the destruction of Government Canyon Bat Cave would eliminate the Government Canyon Bat Cave Meshweaver, an endangered spider known only from that cavern. A localized pollution event would have a similar effect in these habitats.
- Species with narrow habitat requirements: Golden-cheeked Warblers require strips of bark from mature Ashe Juniper trees for nesting

Figure 2.4.
This healthy Coyote is
comfortable in an urban
setting. Photograph
by Frank Schulenburg,
Creative Commons.

material. They construct their nests only in the juniper-studded brushlands that are rapidly being converted to suburban neighborhoods and shopping malls in the Austin–San Antonio corridor. Prior to 1900, as many as 1 million Attwater's Greater Prairie-Chickens occupied the tallgrass prairies of the western Gulf Coastal Plain. Over the intervening years, urban developments, conversion of grasslands to agricultural fields, and the spread of invasive woody plants such as Chinese Tallow gradually eliminated the small, traditionally used display arenas (leks) where males and females gathered to mate. As the specialized habitats required by the warblers and prairie-chickens disappeared, the populations of these species declined.

- Species that are large or require a larger range: The Gray Wolf once roamed widely in the western two-thirds of the state. Because it was large enough to prey on livestock, it was extirpated in Texas by predator control measures. Black Bears were similarly extirpated years ago, but a few recently immigrated from Mexico, especially into the Chisos Mountains of Big Bend National Park, and they are occasionally seen farther south along the Rio Grande. Black Bears from the swamps of Louisiana appear to be reestablishing a population in the dense forests of East Texas, and a few periodically migrate into the state from Oklahoma. Large organisms require more food and larger habitat areas than smaller animals. They also become easy targets.

- Species with limited reproductive potential: This category includes species that produce only a few offspring per breeding season, have long gestation or incubation periods, rear offspring requiring extensive parental care, or require a long period before reaching sexual maturity. For example, the Fin Whale, an endangered species that occasionally inhabits the Gulf of Mexico, exhibits all of these characteristics. After reaching sexual maturity at six to ten years of age, a female gives birth to a single calf, which is suckled for about seven months after birth and remains with the mother for another year. Consequently, female Fin Whales give birth at three-year intervals.

- Species with specialized physical, behavioral, or physiological requirements: Small groups of West Indian Manatees once inhabited the Laguna Madre and other Texas estuaries and river systems. Although they were once exploited as a food source, it is likely that extended winter cold snaps caused their demise in the state. West Indian Manatees are extremely sensitive to cold and die when exposed to water temperatures below 46°F. Although West Indian Manatees occasionally wander into Texas waters, their physiological requirement for warm temperatures prevents them from establishing a permanent population.

- Species with genetic vulnerability: Most species in the mammalian family Canidae, the doglike carnivores, are capable of interbreeding. Specific status among native canids was once maintained by geographic separation and behavioral differences that prevented interbreeding. As Coyotes expanded their range eastward in Texas, interbreeding began to occur between Coyotes and Red Wolves. As a result, the Red Wolf gene pool was in danger of being altered by genetic swamping of Red Wolf genes by Coyote genes. Hybrids became more common, and before long only a few "pure" Red Wolves remained. Before genetically distinct Red Wolves became extinct, 17 Red Wolves lacking Coyote genes were captured and bred in captivity. Genetically pure Red Wolf offspring were released at Alligator River National Wildlife Refuge, North Carolina. Their descendants are again threatened because Coyotes have moved eastward, invading the lowland habitats of the Atlantic Coast.

- Species with international distributions: Protections afforded by the Endangered Species Act (ESA) apply only to the United States, its territories, and its boundary waters. Beyond these zones, species protections are limited and often unenforceable. The endangered sea turtles that nest on Texas beaches spend much of their lives in international waters where they are subject to commercial harvesting for meat and other products, death in trawls lacking Turtle Excluder Devices, and deadly accidental encounters with vessels and marine debris. The Mexican Long-nosed Bat migrates annually to Texas and New Mexico after spending the winter months in Mexican caves. Now protected as an endangered species in both countries, the bats were once killed in winter roosts because of local superstitions.

- Species of economic importance: The populations of several threatened and endangered species declined because of demand for their feathers, fur, or flesh. Ocelots and Jaguars, for example, were hunted for their spotted pelts. Passenger Pigeons were harvested by the millions; pigeon meat was shipped to eastern markets and served in restaurants as cheap food. Similarly, the Eskimo Curlew, which was considered the most abundant shorebird in North America until the end of the nineteenth century, was decimated by relentless hunting until the last remaining birds began to disappear. No reliable sightings of the species have been reported in the last 20 years, and it is probably extinct.

Many species possess several characteristics that prejudice their vulnerability to extinction. Whales are prime examples. These large animals have low reproductive rates, long gestation periods, young that require long periods of parental care, and large home ranges that cross international boundaries and major shipping lanes. After taking several years to reach sexual maturity, they give birth to a single offspring, and as many as three years pass between calvings. In addition, whales were relentlessly harvested for nearly two centuries because their oil, meat, and blubber were highly valued. Many decades of protection will be required for some whale species to reestablish sustainable populations.

Some species possess none of the attributes that should render them vulnerable to extinction, but these unfortunate organisms happen to reside where humans are altering their habitat faster than the species can adapt or emigrate. Fortunately, the United States and many other nations have enacted laws designed to protect conservation-reliant species and the habitats required for their continued existence. The development of species protections in the United States is summarized in the following chapter.

3 PROTECTING AMERICAN SPECIES: A BRIEF HISTORY

BRIAN R. CHAPMAN & WILLIAM I. LUTTERSCHMIDT

This important measure grants the Government both the authority to make early identification of endangered species and the means to act quickly and thoroughly to save them from extinction. — PRESIDENT RICHARD M. NIXON

When President Richard M. Nixon signed the Endangered Species Act (ESA) into law on December 28, 1973, it was widely hailed as one of the most far-sighted statutes ever adopted by any nation. The ESA identified species perched on the brink of extinction, provided a legal means for protecting them and the habitats on which they depended, and developed a framework for conservation measures designed to lessen threats to their survival. Many aspects of the law, however, were not new. Much like the species it was designed to protect, the ESA was the product of long evolutionary processes. Components of the ESA, including some entire sections, were lifted from federal statutes enacted over the course of the previous century and inserted with minor changes into the new law. Each statute that contributed language or regulations to the ESA was originally passed in response to a specific concern at some point in the nation's history. Moreover, the ESA continues to evolve in response to public pressures and political will. What follows is our attempt to provide an abbreviated history of the legal framework that ultimately led to passage of the landmark legislation.

Evolution of Laws Protecting Species and Species Diversity

As settlement progressed westward from the Atlantic Coast to fulfill the nineteenth-century promise of "Manifest Destiny," the pioneers entering the new territories may have been reassured by the seemingly limitless biological resources. Settlers innocently cleared the land and unknowingly overharvested game and furbearing animals without fear of depletion. By the late 1800s, however, most ungulate species had been eliminated from major portions of their original ranges east of the Mississippi River. The eastern subspecies of White-tailed Deer, for example, was almost completely eradicated by 1870. Soon

thereafter, the eastern subspecies of Elk likely became the first distinctly North American animal to experience extinction at the hand of the new Americans.

Large hoofed mammals were not the only organisms to be decimated as the population advanced westward. Hawks were slaughtered to protect domestic fowl (fig. 3.1), and large predators such as the Gray Wolf and Mountain Lion were systematically eliminated because of their perceived threats to humans and domestic animals. Their populations further declined as their main prey—White-tailed Deer and Elk—also began to disappear. In the absence of natural prey, surviving wolves and lions increased predation on domestic animals, resulting in intensified efforts to exterminate them. Populations of the American Beaver and other furbearers were virtually extinct

Figure 3.1. For many years, hunters shot hundreds of hawks as they flew over Hawk Mountain, Pennsylvania, during their annual migration. The site is now a sanctuary where visitors flock to witness the migrations and scientists conduct annual counts. Photograph courtesy of Hawk Mountain Sanctuary.

Figure 3.2. "Martha," the last living Passenger Pigeon. Huge flocks numbering in the millions once darkened the skies of North America. Photograph by Enno Meyer, Creative Commons.

east of the Mississippi by 1830, and many other eastern species, including the Heath Hen, Great Auk, Labrador Duck, and Sea Mink, succumbed to overharvest by 1900. Just after the turn of the century, the last surviving Passenger Pigeon, once the most abundant bird in North America, and the last Carolina Parakeet died within a few years of each other in the Cincinnati Zoo (fig. 3.2). Shortly thereafter, a public announcement was made that the California Condor and several species of waterfowl were nearing extinction. All told, one product of Manifest Destiny was an appalling loss of wildlife.

For many years, the extinctions and decimations of these and many other species either escaped public notice or generated little interest or concern. During this stage in our nation's history, the protection of wildlife was considered a responsibility of state governments. From 1865 to 1900, many states developed a framework for administrative oversight of wildlife issues, but few state laws limited harvests. Most state laws were either toothless or unenforced.

The public finally paid attention to the plight of wildlife when the American Bison had all but disappeared. The noble icon of the American West once dominated the Great Plains in massive herds totaling 20 to 30 million. After just a few decades of unrelenting slaughter, only about 1,000 Bison remained in small groups isolated in remote pockets of the Great Plains and inaccessible valleys of the Rocky Mountains (fig. 3.3). In 1872, Congress took a giant step toward federal protection of wildlife habitat when it created Yellowstone

National Park, which contained one of the last remaining herds of American Bison. Protection of the Bison and other species was difficult, however, because many citizens of Wyoming believed that the wildlife in the park belonged to the state and continued to harvest the animals therein. To resolve this dispute, Congress passed the Yellowstone Park Protection Act of 1894. This law resolved the states' rights issue by establishing a precedent for federal jurisdiction over wildlife—in essence, Yellowstone National Park became the first national wildlife refuge.

Rapid population declines of the Passenger Pigeon, Eskimo Curlew, and several migratory bird species prompted further federal action. For years, market hunters had slaughtered edible birds by the thousands and shipped them by rail to restaurants in eastern cities. It became obvious that states could not respond adequately to wildlife issues, especially those that involved governmental entities outside their jurisdiction. Using the commerce clause of the Constitution as its basis for action, Congress passed the Lacey Act

Figure 3.3. This massive pile of American Bison skulls, gathered in 1892 for shipment to a fertilizer plant, bears witness to the near extermination of the noble beast. Public domain; original photograph is in the Burton Historical Collection at the Detroit Public Library.

Figure 3.4. As they attended their nests, Snowy Egrets and other herons were killed to meet the desire of a fashion industry that used filamentous plume feathers for decorative accents. Photograph by Len Blumin, Creative Commons.

of 1900. This law prohibited the transfer of wildlife, fish, or plants across state lines if the specimens had been taken, sold, possessed, or transported in violation of any other state law. A provision of the statute also authorized the secretary of the interior to restore game animal populations where they were rare or extirpated. In view of its scope, many consider the Lacey Act to be the first federal legislation to confront the problem of species extinctions.

During his terms in office (1901–1909), President Theodore Roosevelt, an avid sportsman and hunter, played an active role in species protection. Using his executive power, Roosevelt established the first official national wildlife refuge at Pelican Island, Florida, in 1903. The refuge protected large rookeries of nesting wading birds and shorebirds, which were being killed for their plumage. The long plume feathers produced by herons and egrets during the breeding season were valued as ornamental decorations on women's attire (fig. 3.4); entire carcasses of small terns were cured to serve as women's headdresses. Plume hunters often shot adult birds on their nest, plucked the desired feathers, and left the nestlings to starve. Before leaving office, Roosevelt extended federal protection to land and wildlife by establishing

55 other wildlife refuges, 18 national monuments, 6 national parks, and over 100 national forests.

Despite the nation's efforts in setting aside public land for protecting wildlife, waterfowl and shorebird populations continued to decline, again demonstrating that states could not effectively prevent overharvesting of wildlife. Groups such as the National Audubon Society, the League of American Sportsmen, and the American Game Protective and Propagation Association pressured the federal government to sign a treaty with Great Britain (acting on behalf of Canada) to provide protection for migratory birds. The Migratory Bird Treaty Act of 1918 ratified the treaty, which had been signed in 1916; amendments ratified similar treaties with Mexico (1936), Japan (1972), and the Soviet Union (1976; now known as Russia). The statute prohibited possessing, hunting, taking, capturing, killing, or selling birds or any of their parts, including feathers, nests, and eggs, without a federally issued permit. The law included provisions for fines and forfeitures for anyone found in violation of the law. Most significantly, the regulation, which was upheld by a Supreme Court decision in 1920, gave the federal government the power to supersede state laws concerning wildlife. Entire phrases from this law and many of its other aspects were incorporated into the ESA five decades later.

Despite the protections afforded by the Lacey Act and the Migratory Bird Treaty Act in the early years of the twentieth century, waterfowl populations continued to decline. The losses were soon attributed to destruction of the birds' breeding habitats in the northern Great Plains of North America. Within this expansive region were millions of prairie potholes—small ponds and wetlands formed in depressions created nearly 10,000 years before, when the last Ice Age glaciers retreated (fig. 3.5). As the Great Plains prairies were plowed under in the conversion of grasslands to agriculture, the prairie potholes were drained and filled. The nesting and foraging habitats for thousands of Mallards, Northern Pintails, Gadwalls, Blue-winged Teal, and many other species simply ceased to exist. Conservationists and duck hunters joined in efforts to prevent further habitat loss and protect the nation's waterfowl. The American Wild Fowlers, an interest group formed to ensure that additional wetlands came under federal protection, led the effort to pass the Migratory Bird Conservation Act of 1929. The statute established the Migratory Bird Conservation Commission and charged it with recommending lands for purchase to the Department of the Interior. The law established federal responsibility for acquiring and protecting wildlife habitats, especially those important to waterfowl production, throughout the United States. In effect, the law established the National Wildlife Refuge System.

Aldo Leopold, who became the chair of the American Game Policy Committee, led efforts in the 1930s to change American attitudes toward wildlife in general and predators in particular. When he became chair of the committee, Leopold himself, the public, most public agencies, and many land managers believed that predator control was the best option for ensuring that healthy populations of game species could be maintained. However, after witnessing the impacts of deer overpopulation on habitats and wildlife after predators had been removed from the Kaibab Plateau, Professor Leopold reexamined the role of predators in a natural system. He was among the first to understand the significance of predators in maintaining a healthy ecosystem.

Congress established a momentous precedent when it passed the Bald and Golden Eagle Protection Act of 1940 (fig. 3.6). The law imposed a nearly complete ban on the taking of Golden and Bald Eagles, including their eggs, nests, and feathers (there were some exceptions for Native Americans). This statute applied nationally and filled a legal gap. The Migratory Bird Conservation Act protected migratory birds, but eagles, which do not migrate, were protected only within the confines of national wildlife refuges. They were being decimated on private lands by ranchers, farmers, and feather collectors. Essentially, the Bald and Golden Eagle Protection Act became the first law to protect species in danger of extinction.

Several significant developments followed during the next three decades. First, President Franklin D. Roosevelt combined the Bureau of Fisheries and the Biological Survey in 1940 to

(top) Figure 3.5. The Prairie Pothole Region extending across five states and three Canadian provinces provides breeding and foraging habitat for waterfowl, shorebirds, and upland birds, including several endangered species. Photograph courtesy of Ducks Unlimited, Inc.
(bottom) Figure 3.6. Without the special protection provided by the Bald and Golden Eagle Protection Act, Bald Eagles and Golden Eagles might have been eliminated in the United States. Photograph by Murray Foubister, Creative Commons.

create the Bureau of Sport Fisheries and Wildlife. This agency would eventually become the US Fish and Wildlife Service. At about the same time, the related scientific disciplines of wildlife management and range management began to blossom both scientifically and professionally, providing professional education to a new generation of biologists. Third, the public began to fully awaken to the importance of the natural environment as the impacts of pollution became more pronounced.

Environmental awareness grew slowly until Rachael Carson published her landmark book, *Silent Spring*, in 1962. The book described the deleterious impacts of organochlorine pesticides on wildlife by imagining a spring without the accompanying melodies of songbirds. Many historians link the beginning of the environmental movement of the 1970s and widespread concern about species preservation to the publication of Carson's bestseller, the debate that followed, and the banning of DDT and other organochlorine pesticides in 1972.

The Initial Endangered Species Acts

An early indication of concern about species preservation was the 1964 publication of the *Redbook*, ostensibly the first official list of animals considered by the federal government to be in danger of extinction. The list of 83 species—all vertebrates—was developed by nine biologists appointed by the Bureau of Sport Fisheries and Wildlife to form the Committee on Rare and Endangered Wildlife Species. Just two years later, Congress passed the Endangered Species Preservation Act of 1966, the first federal statute to address species endangerment. The law, regarded as ineffective by many, directed federal agencies to protect habitat for the species on the previously developed list, but it applied only to animals on national wildlife refuges. Even there, protection was limited to only the extent "practicable," and the law failed to address commerce in endangered species and their parts. It did, however, authorize the establishment of refuges expressly for endangered species conservation.

The Endangered Species Conservation Act of 1969 was developed to improve on the initial statute. Although the new law did not provide protection for species occurring outside federal lands, it expanded the provisions of the Lacey Act to prohibit interstate commerce involving illegally obtained mollusks, crustaceans, amphibians, and reptiles. Two other aspects of this law were especially significant. First, the statute required the secretary of the interior to develop a worldwide list of endangered species and to prohibit their importation. Second, the secretary of the interior

Figure 3.7. Outrage over brutal images of Harp Seal pups being clubbed so their skins could be harvested generated public support for passage of the Marine Mammal Protection Act. Photograph by Matthieu Godbout, Creative Commons.

was directed to enable an international convention on the conservation of species.

As the environmental movement gained momentum, concerns about the decline of most marine mammals, especially whales, became a rallying point. The clubbing of Harp Seal pups in the icy North Atlantic was shown in several television documentaries and news shows, giving rise to perceptions of animal cruelty and focusing additional political pressure on elected officials in Washington (fig. 3.7). Congress responded by passing the Marine Mammal Protection Act of 1972, which prohibited the take of marine mammals. The law broadly defined "take" as the act of hunting, killing, capturing, or harassing any marine mammal as well as any attempt to do these activities. Also banned was the importation of marine mammals and products derived from them. In addition to whales, porpoises, and dolphins, the statute extended protection to the Polar Bear, Walrus, Sea Otter, Dugong, and the seals and marine otters residing in US jurisdictional waters. In subsequent amendments, provisions were made for subsistence harvests by Native Americans in Alaska, permits for scientific research, and a reward system for information leading to convictions for violations of the law.

Many provisions of the Marine Mammal Protection Act were translocated to the ESA.

The Endangered Species Act of 1973

Widely publicized, the first Earth Day (April 22, 1970) played a significant role in educating the public and Congress about the breadth and value of biodiversity in America. Building on the framework provided by the ESAs of 1966 and 1969 and the Marine Mammal Protection Act, Congress passed the Endangered Species Act of 1973. The ESA emphasized that the prevention of extinction was an overriding responsibility of every federal agency, and that federal law superseded state authority. Two categories of imperiled species were designated: "threatened" and "endangered." A threatened species was defined as one that was "likely to become an endangered species within the foreseeable future throughout all or a significant portion of its range." This provision was intended to protect species before they became endangered—defined as "any species which is in danger of extinction throughout all or a significant portion of its range."

Unlike many federal statutes, the ESA of 1973 is neither long nor complicated. It provides a legal framework for (1) identifying species in need of protection, (2) prescribing protection and management measures, (3) defining the habitat essential to the species' survival, and (4) setting measurable goals to periodically assess how well the protection and management measures are working. The law also has "teeth"—it establishes punishments for those who harm listed species. Except for a section of the 1966 law concerning the National Wildlife Refuge System, this act repealed the two earlier ESAs. The ESA of 1973 extended protection to listed plants (but only on public lands) and a larger range of invertebrates, vertebrate subspecies, and groups of animals representing distinct population segments. In other words, if the survival of a significant isolated population appeared to be imperiled, it would be protected as if it were a species; this provision was later amended and now applies only to vertebrates. Animals specifically excluded from protection are insects considered harmful to humans, which in broadest terms applies to agricultural pests and insects that potentially spread disease.

A species gains protection under the ESA after it is listed as endangered or threatened by the US Fish and Wildlife Service (FWS), which is responsible for protecting land animals and freshwater fishes, or the National Marine Fisheries Service (NMFS), which has jurisdiction over sea turtles, marine mammals, and oceanic fishes. The listing processes that relevant federal agencies must follow are prescribed in the Administrative Procedure Act along with several special requirements described in the law itself.

The progression of events by which a species is added to or deleted from the list of threatened or endangered species begins with a petition and is followed by a series of prescribed steps. Movement through the steps is made public through notices published in the *Federal Register* before and after each stage. Either the FWS or the NMFS may initiate the listing process for a species, but petitions can also originate from conservation-oriented organizations or concerned individuals. Once a petition is received, the federal agency receiving the petition must complete an evaluation of the scientific or commercial information contained in the petition within 90 days. If the information does not warrant further investigation, the petition is rejected. If further action might be warranted, however, a more detailed status review is begun.

During the 12 months after a positive finding is announced, the FWS or NMFS must gather all available scientific information and commercial data to evaluate the status of the species. Species experts are consulted and field surveys are often conducted to obtain current population estimates or assessments of threats to the species' continued existence. A six-month extension is permitted if circumstances warrant. Eventually, the agency in charge must make a listing determination after considering five factors: (1) present or threatened damage, modification, or destruction of the species' habitat; (2) reduction of the species' population for commercial, recreational, scientific, or educational purposes; (3) reduction of the population because of disease or predation;

(4) inadequacy of existing protection; and (5) other factors (human caused or natural) that affect the species' continued existence. During this assessment, efforts made by states or counties to protect the species must also be evaluated. The potential economic impact of listing a species, however, may not be considered.

After an initial listing decision is made, the agency must publish a draft finding indicating that listing the species is either "warranted," or "not warranted." After publication of the initial finding, public comments are accepted and evaluated. Based on evaluation of evidence gathered from all sources, including all public comments, the FWS or NMFS then publishes a final rule, which may be subject to judicial review. The final rule establishes the listing status of the species—not warranted, threatened, endangered, or warranted but precluded from listing. The latter indicates that the federal agency determined that the evidence is sufficient to warrant listing the species as threatened or endangered, but it cannot do so because other species have higher listing priorities. Species in this category are designated "candidate" species and must wait their turn for the agencies to conclude the listing processes for species in line ahead of them. These at-risk species do not benefit from federal protection until they are eventually listed as either threatened or endangered.

Once an animal species is listed, any activity that causes harm to the species or its habitat is prohibited. The law stipulates that it is unlawful for any person subject to the jurisdiction of the United States to "take" a listed species within the United States, in the territorial waters of the United States, or on the high seas. The word "take" is broadly defined—"to harass, harm, pursue, hunt, shoot, wound, kill, trap, capture, or collect, or to attempt to engage in any such conduct." Federal courts later interpreted "harm" to include any habitat modification or degradation that could result in extinction by significantly impairing essential behavioral patterns, such as feeding, breeding, or sheltering. Threatened and endangered plants receive protection from "take" only on federal lands

When it is "prudent and determinable," a listing determination is accompanied or followed closely by designation of critical habitat for the species. Critical habitat encompasses geographic areas that contain the physical or biological features essential to conserving the species and are in need of special protection. Any area essential to the conservation of the species may be included as critical habitat even though the species may not occupy the area at the time of listing.

Unlike the listing determination, however, designation of critical habitat requires that the listing agency consider economic, national security, and other impacts before reaching a final decision on area and extent. Unless not designating an area as critical habitat may lead to the extinction of a species, the agency can exclude portions of a necessary habitat if the economic benefits of excluding it outweigh the benefits of including it. Determinations of critical habitat usually generate much interest during the public comment period following the draft description appearing in the *Federal Register*. Once critical habitat has been defined, federal agencies or federally funded permit activities must seek permit approval to conduct any activity that might modify or destroy the natural resources supporting the species. Individual landowners are not required to obtain a federal permit to modify their property, but they are constrained by the prohibition of "take" and "harm."

Protection of at-risk components of the nation's biodiversity is only one goal of the ESA. The ultimate goal is to allow the listed species to persist in the wild by reducing or eliminating the factors causing its decline. After a species is listed, biologists employed by the FWS or NMFS work with species experts from academia and multiple other stakeholders to develop a recovery plan. The recovery plan describes the efforts believed necessary to reverse the population decline, establishes measurable population goals that would indicate progress or success, and creates a timeline for the actions to take place. The recovery plan serves only as a "road map" for species recovery—the management activities prescribed within it are not mandatory and no funds are committed with its publication. Nonetheless, federal and state agencies, nongovernmental organizations, and private citizens often participate in supporting the prescribed recovery efforts. Every five years after a species is listed, a status review must be conducted to determine whether the population has recovered enough to allow a change of listing status or a delisting.

From its inception, critics of the ESA have argued that preventing economic development to ensure the survival of one species among the millions of species in the world is pointless. The obvious underlying concern is that the protection of listed species and the designation of critical habitat will limit the development of property and business. It is extremely rare, however, that a proposed project is terminated because it would cause jeopardy to a listed species. Both the FWS and NMFS provide consultation in which they offer "reasonable and prudent alternatives" describing how the proposed activity could be modified to avoid jeopardy to the species or habitat.

Effective federal-state partnerships are critical to species conservation efforts. The ESA encourages states to maintain conservation programs for threatened and endangered species, and the federal government provides funding to promote participation. Most states, including Texas, maintain their own list of threatened and endangered species, and some have laws and regulations that are more restrictive than the ESA in granting permits.

The ESA also prohibits trade in specimens listed on any of the three appendices within the Convention on International Trade in Endangered Species of Wild Fauna and Flora (CITES). All species protected as threatened or endangered by the ESA as well as those designated by CITES and other countries are listed in the CITES appendices. Under the terms of CITES, all listed species receive the same protection from commercial exploitation. For example, importation of ivory or products made from the ivory of African Elephants is prohibited, as is the importation of wood or wood products derived from certain rare trees. A citizen of the United States might purchase an ivory or wood product in a country that does not subscribe to CITES or where wildlife protections are lax, but the product cannot be legally imported into the United States or any of its territories. A person

bearing such goods would be forced at the port of entry (including airports) to surrender the items to customs authorities and might be subject to fines or imprisonment.

Private Landowner Obligations and Options under the ESA

When a threatened or endangered species inhabits their property, private landowners are obliged to protect the species and maintain the habitat conditions necessary for its continued existence. In some circumstances, a landowner may incur significant personal expense to this end. Approximately two-thirds of all federally listed species occupy private landholdings to some degree, and some species have much of their remaining habitat on private land. To ease the burden borne by these citizens, the federal agencies (principally the FWS) have developed several options designed to protect the interests of landowners while supporting management activities that benefit listed and other rare species. Some of these are also available to corporations, Native American tribes, states, and counties that want to develop property inhabited by listed species. These options include the following:

HABITAT CONSERVATION PLAN. This option is for a landowner who wants to develop a private property that is inhabited by a listed or candidate species where the proposed activity might "take" some individuals of the species. The FWS will consider allowing the development and may issue an incidental take permit if the landowner develops an approved habitat conservation plan (HCP). The plan must specify ways that the anticipated take of the listed species will be minimized and mitigated. The goal is to improve the conservation value of the remaining land in a manner that contributes to the recovery of the species as a whole. As a part of the HCP, the landowner must demonstrate that there is adequate funding to support the plan. Although incidental take permits have expiration dates, the mitigation required may exist in perpetuity. Because "take" is involved, a violation of the terms of the incidental take permit is considered an unlawful act under the ESA.

SAFE HARBOR AGREEMENT. Whereas an HCP and a related incidental take permit are an option for landowners who could potentially harm a listed species, a safe harbor agreement (SHA) protects landowners who voluntarily act to improve the habitat for a listed species occurring on their property. Before an SHA can be established, the federal agency cooperates with the landowner to establish a baseline population estimate for the listed species on the property. With an SHA in hand, the landowner can make habitat improvements without fear that the federal agency might impose future additional regulations. When the SHA expires, the landowner can alter the property in any manner as long as the baseline population remains. If the habitat improvements are successful and result in a population increase, the landowner will be issued an incidental take permit, allowing any legal activity that might result in "take" of the listed species as long as the agreed-on baseline population is maintained.

CANDIDATE CONSERVATION AGREEMENT. It is usually more difficult to conserve species after they are listed than before they are in danger of extinction. A candidate conservation agreement (CCA) is a voluntary covenant between federal or state agencies and one or more other parties to reduce or eliminate threats to candidate or other at-risk species. The CCA requires that all parties to the agreement work with the FWS to design conservation measures, implement the plan, and monitor their effectiveness.

CANDIDATE CONSERVATION AGREEMENT WITH ASSURANCES. Private landowners may volunteer to participate in a candidate conservation agreement with assurances (CCAA). Under a CCAA, landowners agree to cooperate with the FWS on efforts to conserve candidates and other at-risk species on their property. In return for their investments of time and money, landowners receive assurance that they will not be subject to additional restrictions beyond those specified in the agreement should the species become listed under the ESA. They also receive an enhancement of survival permit, which allows incidental take

if the management activities specified in the agreement harm the species.

CONSERVATION BANKS. A conservation bank is a permanently protected property that is conserved and managed for species that are threatened, endangered, candidates for listing, or otherwise at risk. These properties offset adverse impacts that have occurred elsewhere. When a landowner enters a property into a conservation bank, the FWS awards the owner a specified number of habitat credits. The landowner may sell these credits to others who need to mitigate the adverse impacts their proposed project may cause to the same species elsewhere. Species benefit from conservation banks because a large habitat is conserved. This approach to mitigation reduces the number of small, isolated, and unsustainable preserves that often lose their habitat value over time.

Species Protection Successes and Disappointments

The ultimate goal of the ESA is to "recover" species so that their populations are stable and protection under the ESA is no longer needed. There are many success stories. For example, the number of Brown Pelican (fig. 3.8) nests in Texas increased from 8 in 1970 to over 10,000 by 1995. The American Alligator and Peregrine Falcon, which were once listed as endangered species, were removed from the list decades ago as a result of protection and management. The Louisiana Black Bear was also recently delisted. Although the Whooping Crane remains on the list, its population has rebounded from 54 birds when it was listed as an endangered species in 1967, to over 500 today. The ESA is the strongest and most effective law passed by any nation for protecting biodiversity.

Like any comprehensive federal law, the Endangered Species Act has been reauthorized and amended numerous times during its history. Most of the changes have allowed the responsible federal agencies greater latitude in working with public and private interests to minimize economic impacts while still providing essential protections and management to allow species recovery.

Figure 3.8. The Brown Pelican (left) and American Alligator (right) were once listed as endangered but were later removed from the list of protected species because of successful protection and management strategies. Photographs by Frank Schulenberg, Creative Commons (left), and Steve Hillibrand, US Fish and Wildlife Service (right).

Nevertheless, political forces sometimes exert a stronger influence than the biological information that supports protection of a species under the ESA. A recent example involves the court-ordered delisting of the Lesser Prairie-Chicken, which may ultimately suffer extinction as a consequence of political pressure.

The Lesser Prairie-Chicken was considered for inclusion in the first list of species considered for federal protection under the Endangered Species Act, but it was not listed. The species was ultimately listed as a candidate species in 1998 because of concerns about environmental disturbances and habitat loss associated with energy development. In response, the Lesser Prairie Chicken (LPC) Interstate Working Group (Colorado, Kansas, Oklahoma, Texas, and New Mexico) of the Western Association of Fish and Wildlife Agencies (WAFWA) developed a range-wide conservation plan to increase the population size of the species and encourage conservation via landowner-agency partnerships. The goal of the plan was to provide voluntary conservation efforts in the hope that listing could be avoided. The plan also included a mechanism for compliance with the ESA if the species was listed.

In 2014, the species was listed as threatened, but included in the listing was a special rule that endorsed implementation of the LPC Range-Wide Conservation Plan to address the conservation needs of the species. This was significant because

it provided regulatory certainty to enrolled landowners and industry participants. For enrolled participants, incidental take of the species was allowed if they were operating in compliance with the LPC Range-Wide Conservation Plan. Despite the provisions of the special rule, five lawsuits, including one by the Permian Basin Petroleum Association and four by counties in New Mexico, were filed against the listing decision and the listing process.

The listing of the Lesser Prairie-Chicken was vacated by the US District Court for the Western District of Texas (Midland, Texas) on 1 September 2015. However, in November 2016, the US Fish and Wildlife Service initiated a 90-day status review of the Lesser Prairie-Chicken in response to a petition from WildEarth Guardians, the Center for Biological Diversity, and Defenders of Wildlife. The petition requested listing of three of its distinct population segments as endangered and additionally requested that the Sand Sage and Shinnery Oak prairie population segments be emergency listed as endangered under the ESA. The US Fish and Wildlife Service review resulted in a "finding of substantial scientific or commercial information indicating that the petitioned actions may be warranted." Essentially, this was a request for information. The proposed rule was posted on 30 November 2016, and the comment period ended on 30 January 2017. Whether the species will ultimately be listed is unknown. In the meantime, conservation actions associated with the LPC Range-Wide Conservation Plan continue.

In the past, presidents have temporarily blocked the addition of species to the ESA threatened and endangered list, but those attempts have been overturned by federal courts. The recent ruling overturning the listing of the Lesser Prairie-Chicken by a federal judge was apparently based on economic concerns rather than biological data. Efforts to protect the nation's biological heritage have certainly been complicated by the recent judicial decision.

The ESA was a giant step forward in the protection of biodiversity, but it was not lacking in critics. Enactment of the ESA created widespread fears within the business community about its potential impact on commerce, property development, and the economy. Private landowners were also concerned that any threatened and endangered species occupying their property would not only restrict how they could use their land but would also impose requirements to provide (at their expense) the habitat management required under the law to maintain the proper habitat conditions. To accommodate these real concerns, the ESA has undergone numerous modifications through the years to allow businesses and landowners many viable options. As a result, only a handful of development projects have been completely disallowed, and with the assistance of citizens, corporations, and private nongovernmental organizations, many species once on the brink of extinction have recovered.

When the last Passenger Pigeon, a female named Martha, died alone in her cage at the Cincinnati Zoo, future generations lost forever the chance to witness flocks of this colorful bird darkening the skies (see fig. 3.2). Martha died in 1914, the last Carolina Parakeet died in 1918, and the last Heath Hen followed in 1932. Museum specimens and a few dim photographs have survived the years, but we will never know the thrill of seeing them in the wild. Fortunately, the recovery of many species—such as the Whooping Crane, once nearing extinction with fewer than 30 remaining individuals—can be attributed to the ESA. Thousands of people now flock to the Texas

coast each winter to see America's tallest bird, the Whooping Crane, dance in the marshes as mating rituals begin anew. Sadly, some species continue to disappear, but the ESA represents our best effort to allow future generations of hunters, hikers, campers, and outdoor enthusiasts to see and enjoy this nation's richly diverse flora and fauna. Without this significant law, the United States would be a much less interesting place.

Protection of the State's Imperiled Fauna

After the passage of the Endangered Species Act of 1973, Texas was among the first states to establish a state list of threatened and endangered species. A group of biologists and concerned conservation-minded citizens who had been meeting informally since 1971 founded the Texas Organization for Endangered Species (TOES) in 1973. This nonprofit organization grew rapidly and spearheaded initial efforts to inventory all available information on the rare or imperiled flora and fauna of the state. TOES was instrumental in persuading the state legislature to pass three state laws protecting imperiled species—the State Endangered Species Act, the Native Plant Protection Act, and the Special Nongame and Endangered Species Act.

In many respects, the State Endangered Species Act mimics the federal Endangered Species Act—it enacts similar protections and imposes comparable restrictions. Texas law prohibits the taking, possession, transportation, or sale of any animal designated by the state as threatened or endangered without a state-issued permit. A state listing restricts human activities that would alter habitat and render the organism more susceptible to extinction. Failure to comply with the state law can result in fines, imprisonment, and confiscation of possessions involved with the transgression. The state law, codified in chapter 68 of the Texas Parks and Wildlife Code, also prescribes a process for listing state species and provides a mechanism for adding newly listed federal species to the state list.

Until the Texas law was altered in 2011, any citizen or group could submit a petition containing substantial evidence to support listing of a state species not protected by the federal Endangered Species Act. The petition was submitted to the director of the Texas Parks and

Wildlife Department (TPWD). If TPWD scientists subsequently confirmed that the petition was worthwhile, the department would hold well-advertised public hearings to allow additional input. Based on the findings of the hearings, the director could file an order with the secretary of state to add a species to the state list. The addition took effect immediately upon filing.

When TOES was active (1973–1997), it originated many petitions for listing state species. In 1997, the organization ceased to function independently when it became a section of the Texas Academy of Science. Soon thereafter, TOES lost its identity, membership, and focus. Other conservation organizations and the Texas Parks and Wildlife Department assumed greater roles in petitioning to add species to the state list of threatened and endangered species.

In 1983, The Nature Conservancy and the Texas General Land Office combined their resources to initiate the Texas Natural Heritage Program. Biologists and technicians working with this program developed a database containing information on the location, number of populations, threats to survival, and other criteria for the animals and plants of the state. Information in the database was invaluable for listing decisions or judgments by private and public conservation organizations regarding potential land purchases for parks, preserves, or wildlife management areas. Political pressure from opponents of endangered species protection resulted in the termination of the formal Natural Heritage Inventory Program in the mid-1990s.

During the next 16 years, TPWD biologists continued to obtain and catalog distribution and abundance data on the state's fauna. Their efforts were hampered by reductions in state funding for species inventory efforts and a new state law that restricted the ability of state agency biologists to work on private property. The situation deteriorated abruptly in 2011 when the Texas legislature transferred responsibility for endangered species conservation from the TPWD to the Texas Comptroller of Public Accounts. The comptroller's office serves as the state's treasurer, chief tax collector, revenue estimator, and accountant. The comptroller at the time likened endangered species to "incoming Scud missiles" that threatened to damage the Texas economy.

The comptroller acted quickly to prevent species from being listed for protection by the US Fish and Wildlife Service. For example, the comptroller directed the Texas Habitat Conservation Foundation, a nonprofit organization funded by the Texas Oil and Gas Association, to develop a habitat conservation plan for the Dunes Sagebrush Lizard. The small lizard's distributional range was centered in the Permian Basin, a major oil and gas field. The habitat conservation plan paid landowners with funds from the comptroller's office to preserve the lizard's habitat. Within just a few years, however, a network of well pads and associated roads permeated the protected area, as shown in satellite imagery published by Defenders of Wildlife, a nonprofit conservation group. In 2015, a newly elected comptroller found that the lizard conservation program had delivered little of the preservation and restoration it had promised.

In an attempt to restore confidence in the conservation of the state's biological resources, Senator Kel Seliger convinced the legislature to pass a bill shifting endangered species oversight to a special task force. Although the task force would have had representatives from Texas A&M University, the TPWD, and other state agencies, Governor Rick Perry vetoed the bill. That same year, the legislature allocated $5 million to fund research on species under consideration for listing by the federal government. The law specified that the research was to be conducted by "state public universities with demonstrated experience in species or habitat research, evaluation, and analysis." The comptroller employed a biologist who had worked with the Department of Justice to fight endangered species lawsuits. Controversies developed between the comptroller's office, private consulting firms, and university researchers, leaving questions in the minds of many about the ability of the comptroller to deliver sound science.

The TPWD continues to maintain the list of threatened and endangered species for the state of Texas. The state list includes only the species of plants and animals that occur within the borders of Texas, including many species not on the federal list. Examples of state-listed animal

Figure 3.9. Few populations of the Texas Horned Lizard remain, and the species is included on the Texas list of threatened and endangered species. Photograph by Ben Goodwyn, Creative Commons.

species not included on the federal list are the Gray Hawk, Texas Horned Lizard (fig. 3.9), and Texas Kangaroo Rat. The Texas list currently contains 179 animal species, including 43 invertebrates, 34 fish, 17 amphibians, 25 reptiles, 32 birds, and 28 mammals. Although all of these are rare to some degree within the state, not all are imperiled. Some species represent organisms living at the periphery of a distributional range extending into an adjacent state or Mexico where the population is stable. Oddly, however, some federally protected species that occur exclusively within the state do not appear on the Texas list of threatened and endangered species. Examples include the Georgetown Salamander, Red Knot (*rufa* subspecies), and Western Yellow-billed Cuckoo.

The TPWD helps Texas landowners implement land management practices that benefit imperiled species and rare habitat types by offering grants through the Landowner Incentive Program. In addition, free technical assistance programs are offered by the TPWD, the Texas A&M AgriLife Extension Service, and Texas A&M Forest Service to landowners or land managers desiring to improve wildlife habitat on their property. The US Fish and Wildlife Service and the Natural Resources Conservation Service are also available for consultation. The collaboration between Texas citizens and state and federal agencies continues to set a national example to the benefit of the state's rich biodiversity.

4 THE NATURAL REGIONS OF TEXAS

BRIAN R. CHAPMAN & WILLIAM I. LUTTERSCHMIDT

Texas is a place where major life zones intersect. The flora and fauna include elements characteristic of the southeastern forests, the southern tropics, the southwestern deserts, the western mountains, and the northern Great Plains, as well as marine life associated with the extensive coast. — DAVID J. SCHMIDLY

As dawn breaks, a thick blanket of lofty trees secures the landscape and limits the view of horizons on either side of the road. Welcome to the Piney Woods of East Texas, the starting point of a westward trip on Interstate 20 across the breadth of Texas. Within an hour, the sun has climbed high enough to shine brightly in the rearview mirror. As you concentrate on the road ahead, the forest gives way to grassy meadows and tilled blacklands. On the approach to Dallas, the thick hardwood swaths isolating verdant pastures in the Blackland Prairie region mimic the dense clots of highway traffic. Beyond the brick canyons of Dallas and Fort Worth, the soil takes on a reddish hue and the vegetation becomes increasingly scrubby. Clumps of dark green junipers studding the undulating terrain give way to the lacy light greens of mesquites. Now, with the sun high overhead, dust devils whirl and distant mirages appear as welcoming lakes on the horizon. Within a few more hours, you traverse a region of sandy dunes festooned with willows and low, shrub-like oaks. Beyond these dunes, with the sun shining directly in your eyes, you glimpse the soft purple outlines of mountains emerging from the hazy vastness of the Trans-Pecos. Hardpan soils coated with a shallow layer of pebbles support a network of regularly spaced, olive-drab bushes on each side of the highway. Shadows gradually lengthen as the sun sinks in the western sky, but you still face at least three more hours of driving through the Chihuahuan Desert before you exit the western border of the Lone Star State.

No matter which of the major highways you use to cross Texas from east to west or north to south, you will only begin to experience the full flavor of the state. Your drive along Interstate 20 completely missed the High Plains, the South Texas Brushlands, the Hill Country of the Edwards Plateau, and the Gulf Coastal Plains. The major north-south route across Texas, Interstate 35, similarly bypasses the Trans-Pecos, the Rolling Plains, and the Piney Woods.

Many authors have attempted the daunting task of describing the exceptional habitat diversity within this huge state. Perhaps the most interesting portrayal was published in a recent issue of *Texas Parks and Wildlife* magazine. Two of the most accomplished birders in Texas visited the eight corners—the locations where the state's boundaries make a sharp turn in direction. The habitat at each crook in the geopolitical boundary was highlighted with a short description and a brief listing of the bird species unique to the region. Gambel's Quail and Greater Roadrunner, denizens of the "El Paso Corner," where Creosotebush and Lechuguilla dominate the dry, sandy desert (fig. 4.1, top), were nowhere to be found in the marshes and woody tangles of the "Southeast Corner," where the Sabine River meets the Gulf (fig. 4.1, bottom). The other far-flung corners, harboring a subtropical woodland, prairie, pine forest, and parched desert, differed just as significantly in habitat and avian species composition.

The 268,820 square miles contained within its eight corners make Texas second only to Alaska in size. Any geographic area of that dimension is bound to encompass considerable variation in climate, soils, topography, and their dependent flora and fauna. Its lower-middle position on the continent adds another interesting dimension— it is situated at an ecological crossroads where biological elements of the west meet those of the east and northern elements overlap with those representing the south. Furthermore, the relatively warm waters of the Gulf of Mexico along the state's 624-mile coastline provide a moderating influence on temperatures during the winter and a fairly reliable source of rainfall, especially to the southeastern portion of the state.

Average annual rainfall across the state varies

Figure 4.1. A comparison of habitats typical of the El Paso Corner (top) and the Southeastern Corner (bottom) demonstrates how significantly the "corners" of the state differ. Photographs by Brian R. Chapman.

from less than 8 inches in the Chihuahuan Desert in the extreme west to 30 inches in the central region, to over 55 inches in the swamps along the eastern border. Most people who traverse the state on interstate highways would characterize it as "flat," but more than 91 mountain peaks scrape the clouds over a mile above sea level (fig. 4.2). Fifteen major rivers and more than 3,800 named streams meander through leafy riparian zones, faltering briefly in any of the state's 212 reservoirs along their route before flushing coastal estuaries with freshwater. These conditions endow Texas with the greatest floral and faunal diversity of any state except California. The biological crossroads is home to more than 5,000 species of plants, 145 species of native terrestrial mammals, more than 30 marine mammals, over 630 species of birds, 225 species of reptiles and amphibians, 650 species of freshwater fishes, and nearly 550 species of saltwater fishes in Texas Gulf waters. The total count of invertebrate species inhabiting the state remains unknown, but the number must be staggering.

Through the years, many attempts have been made to delineate the natural regions (or ecoregions) of Texas. More than a century of biological surveys conducted by dedicated biologists often working in dangerous and difficult conditions eventually revealed the distributional patterns of plants and animals in the state. The flora and fauna of any region are not randomly distributed but demonstrate affinities to specific soils, elevations, and climates—the conditions collectively known as "habitat." The adaptations of some species such as the White-tailed Deer or Hispid Cotton Rat permit them to occur widely, whereas others habituated to a relatively narrow set of ecological conditions must exist in a far more restricted range. Over large areas where broad similarities in biotic and environmental factors exist, the plants and animals form recognizable assemblages often defined by dominant vegetation types.

Each of the 10 generally recognized natural regions of Texas (fig. 4.3) is described briefly in the following summaries. The nearshore environment of the Gulf of Mexico is characterized in chapter ten.

Figure 4.2. Emory Peak, 7,825 feet above sea level in the Chisos Mountains of Big Bend National Park, is encircled by a "sky island" forest of Ponderosa and Pinyon Pines. Photograph by Brian R. Chapman.

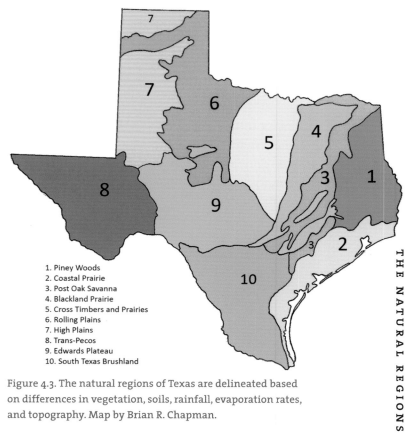

1. Piney Woods
2. Coastal Prairie
3. Post Oak Savanna
4. Blackland Prairie
5. Cross Timbers and Prairies
6. Rolling Plains
7. High Plains
8. Trans-Pecos
9. Edwards Plateau
10. South Texas Brushland

Figure 4.3. The natural regions of Texas are delineated based on differences in vegetation, soils, rainfall, evaporation rates, and topography. Map by Brian R. Chapman.

PINEY WOODS. East Texas is blanketed by extensive stands of pine and pine-hardwood forests on sandy, acidic soils (fig. 4.4, left). Average annual rainfall in the region varies from 40 to 58 inches, and the humidity and temperatures are high. Gently rolling topography with elevations varying from 200 to 700 feet characterizes the region. The Texas Piney Woods are the western-most extent of the pine-hardwood forests that extend across the Southeast to the Atlantic and northward into eastern Oklahoma, Arkansas, and other states. East Texas forests also include other interesting natural communities, including Baldcypress swamps, bogs, wetland savannas, and flatwoods ponds (fig. 4.4, right). Clearings for small ranches and farms interrupt the forest. Most of the original Longleaf Pine forest in the southeastern portion of the region was harvested prior to 1940, leaving only small remnant stands. Losses of old-growth timber led to the extinction of the Ivory-billed Woodpecker and the decline of the Red-cockaded Woodpecker and Louisiana Pinesnake.

POST OAK SAVANNA. The original savannas lying to the west of the Piney Woods were dominated by chest-high grasses such as Little Bluestem and Brownseed Paspalum surrounding patchy woodlands of Post Oak and other trees (fig. 4.5). Annual rainfall varies from 28 to 45 inches, increasing from the southwest part of the region to the northeast. Such annual rainfall ordinarily supports a forest, but the frequency of natural fires every 5 to 10 years and periodic droughts restrict tree and shrub growth. Land clearing and fire suppression have allowed smaller trees and dense expanses of shrubby vegetation to invade savanna habitats. Peat bogs and marshes occur where the water table is near the surface, for example in Gonzales County.

BLACKLAND PRAIRIE. The most common soil of the Blackland Prairie region is often called "black gumbo" because it is a uniformly dark-colored alkaline clay that is sticky when moist (fig. 4.6). Gray, acidic sandy loams also occur in some areas. Rainfall varies from about 27 to 44 inches across the region, increasing from southwest to northeast. Before settlement, a true tallgrass prairie dominated by Little Bluestem, Big Bluestem, Indiangrass, and a variety of wildflowers thrived in the region's deep, nutrient-rich soils. Nowadays, only about 0.1 percent of the original prairie has escaped conversion to other land uses, such as farming and urban development, or invasions by woody species such as mesquite or nonnative grasses like King Ranch Bluestem, Johnsongrass, or Tall Fescue. The few enduring remnants are managed mostly by private landowners, but they are vanishing rapidly except for those protected by organizations like The Nature Conservancy.

CROSS TIMBERS AND PRAIRIES. The Grand Prairie or Osage Plains, as this region is sometimes called, is part of the southernmost region of tallgrass prairies in the United States. Strips of dense oak-hickory forests, the "Cross Timbers," transect the region north to south, separated by expanses of prairie. The Cross Timbers were barriers to early travelers moving west (fig. 4.7). The topography consists of gently rolling hills with elevations ranging from 450 to 1,600 feet. Rainfall peaks in May or June, with annual amounts varying between 28 and 42 inches. Post and Blackjack Oak are common in upland habitats, and Sugar Hackberry and Pecan form riparian forests. Grassy bogs and marshes occur where the water table is near the surface in Comanche County. The region is known for large cattle ranches and pecan groves.

ROLLING PLAINS. Together with the High Plains, the Rolling Plains region represents the southern extent of the Great Plains covering the central United States. Bordered on the west by the Caprock Escarpment, to the south by the Edwards Plateau, and in the east by the Western Cross Timbers, the original prairie grasslands of the Rolling Plains consisted of tallgrass and midgrass species of bluestems and gramas. Rainfall peaks occur in May and September, but the total rain varies between 20 and 30 inches. Reddish, slightly alkaline soils composed of sands near streams or tight clays and shales in upland sites characterize the region (fig. 4.8). Elevations vary from 800 to 3,000 feet

Figure 4.5. Scattered clusters of Post Oaks and other hardwoods stud the grasslands of the Post Oak region. Photograph by Brian R. Chapman.

Figure 4.6. Encroachment by pocket gophers and a few other fossorial species is restricted by the black clay soils of the Blackland Prairie, but the "gumbo" is no match for the tractor and plow. Photograph by Brian R. Chapman.

above sea level. Three subregions, the Mesquite Plains, the Escarpment Breaks, and the Canadian Breaks, are easily distinguished by location and topography. The largest expanse of the Rolling Plains today, the Mesquite Plains, is best described as mesquite-shortgrass savanna where oaks, junipers, acacias, and prickly pears are important secondary elements. Steep cliffs and canyon slopes occurring below the Caprock Escarpment constitute an ecotone between the High Plains and the Rolling Plains. Grasslands peppered by brushy

Figure 4.7. Wide swaths of impenetrable forest crossing verdant prairies hindered westward travel by early Texas pioneers. Photograph by Brian R. Chapman.

Figure 4.8. The reddish soils of the Rolling Plains are evident where a gully slices through the grassy plain. The caprock escarpment of the High Plains looms in the background. Photograph by Brian R. Chapman.

Figure 4.9. The Canadian River cuts across the High Plains of the Texas Panhandle and supports the Canadian Breaks and habitat typical of the Rolling Plains. Photograph courtesy of the National Park Service.

species typify the escarpment habitat, which also extends eastward over 60 miles from Palo Duro Canyon in extensively serrated topography. The Canadian Breaks subregion dissects the High Plains caprock on either side of the Canadian River as it flows through the northern Texas Panhandle (fig. 4.9). The vegetation of the Canadian Breaks is similar to that in the Escarpment Breaks subregion.

HIGH PLAINS. The High Plains region occupies a relatively high, level plateau with elevations ranging from 3,000 to 4,500 feet. The region receives between 15 and 21 inches of rain each year, but the rainfall is not distributed evenly throughout the year, and extensive droughts occur periodically. The High Plains is devoid of naturally occurring trees except for Eastern Cottonwoods scattered along the few watercourses. The vegetation varies with soil type but is generally classified as mixed-grass prairie or shortgrass prairie (fig. 4.10). Buffalograss is the dominant species in the extensive shortgrass prairie. Saucer-shaped depressions called playa lakes hold water in the otherwise flat landscape and provide aquatic resources to migrating waterfowl and shorebirds. Farming occurs throughout the region, supported by irrigation from wells tapping an extensive underground aquifer, but large expanses of rangeland devoted to cattle production remain. Near the top of the Texas Panhandle, the canyon system called the Canadian River Breaks divides the High Plains into northern and southern sections. Mesquite and yucca have invaded portions of the region.

EDWARDS PLATEAU. Commonly known as the "Hill Country," the Edwards Plateau in west-central Texas varies in elevation from about 500 to 3,000 feet (fig. 4.11). The eastern margin of the Edwards Plateau rises abruptly from the Coastal Prairie along the Balcones Escarpment, a crescent-shaped geologic feature extending from Del Rio to Dallas. The Pecos River precisely defines the western boundary of the Edwards Plateau, but the northern border merges gently into the prairies. Annual rainfall ranges from 15 to 35 inches, increasing from west to east. The soils are typically shallow and overlie thick layers of limestone honeycombed by caves, sinkholes, and other karst features. The underground habitats have existed for millennia, allowing many species to fully adapt to the darkness. Many cave-adapted animals lacking pigment and functional eyes are restricted to single cave systems and are listed as threatened or endangered. Springs and seeps emanating from the Edwards Aquifer underlying much of the region contribute to the flow of the many crystal-clear rivers and streams flowing through the canyons along the Balcones Escarpment. Invertebrates, fish, and salamanders restricted to isolated spring outflows face an uncertain future as human demands for water from the aquifer exceed the rate of recharge. Much of the plateau is covered by oak mottes, junipers, and a complex of grasses and shrubs. Granite domes, the largest of which is known as Enchanted Rock, rise out of a central area known as the Llano Uplift.

TRANS-PECOS. The most geologically complex region in Texas, the Trans-Pecos, occupies the area west of the Pecos River. Elevations here range from 2,500 to more than 8,500 feet. The soils of the region generally represent outwash materials from the limestone or volcanic rocks forming the mountains. Although most of the region experiences less than 15 inches of rainfall annually, as well as hot temperatures and high evaporation rates, the higher mountain ranges usually receive rainfall amounts exceeding 20 inches. Rainwater often drains off rapidly, creating occasional flash floods. Some, however, trickles down through porous limestone layers or volcanic pores to recharge aquifers or emerge as springs or seeps. Some fishes and invertebrates have existed in these aquatic habitats since the Pleistocene, when water was more prevalent, and have become unique species. Now, their isolated habitats are vulnerable to drying or modification, rendering them susceptible to extinction. West of the Pecos River where moisture is a rare commodity and evaporation rates are high, desert vegetation including grasses, Lechuguilla, Creosotebush, Ocotillo, yuccas, and cacti dominates the landscape (fig. 4.12). At higher elevations and in protected montane canyons, oak forests, Pinyon Pine, and Ponderosa Pine contribute to the

Figure 4.10. Shortgrass prairie invaded by yucca and mesquite characterizes most of the High Plains. Photograph by Leaflet, Creative Commons.

Figure 4.11. Deeply incised by rivers and streams, the Edwards Plateau is affectionately known as the "Texas Hill Country." Photograph by Mark Nelson, Creative Commons.

biological diversity of the region. About 1 out of every 12 plant species occurs nowhere else, and the Trans-Pecos region also contains many endemic reptiles.

SOUTH TEXAS BRUSHLAND. The "Brush Country" east of the Rio Grande is characterized by dense associations of thorny shrubs and trees (fig. 4.13). The region was once dominated by extensive grasslands, and an extensive palm forest and scattered tracts of subtropical woodlands occupied the Rio Grande Valley. Average annual rainfall increases west to east from 20 to 32 inches, but rains are seasonal, occurring primarily in the

Figure 4.12. A recent rainfall sparked emergence of colorful but short-lived flowering plants in an otherwise dry habitat. The Chisos Mountains loom in the background. Photograph by Brian R. Chapman.

Figure 4.13. Wildflowers abound when spring arrives in the South Texas Brushland. Photograph by Brian R. Chapman.

Figure 4.14. Small patches of Live Oaks called "oak mottes" provide feeding and resting areas for migratory songbirds in the Coastal Prairie. Photograph by Brian R. Chapman.

spring and fall. Temperature and humidity remain high throughout the year, and evaporation rates often exceed rainfall. If any natural region could be highlighted as a "biological crossroads," the Brush Country would be it. The distributional ranges of many species overlap here—semitropical species extend their range northward from Mexico to coexist with desert species from the Trans-Pecos, and woodland and grassland species from the Edwards Plateau, Coastal Prairie, and High Plains also occur here. Consequently, the South Texas Brushland is a haven for many rare plants and animals.

COASTAL PRAIRIE. The nearly level plain bordering the Gulf Coast rarely exceeds an elevation of 200 feet. Rainfall is distributed evenly through the year but varies across the region, from about 20 inches near the Rio Grande to slightly over 60 inches at the Sabine River. After rains, the land drains slowly, but many rivers and streams slice the prairie to flush coastal bays and estuaries. Prairie soils are mostly clays or sandy loams, but windblown sands accumulate in dunes near the coast, especially in an area south of Baffin Bay known as the Sand Plains or Sand Sheet. The dominant vegetation of the region once consisted of tall grasses and oak mottes, which lined portions of the coast (fig. 4.14). The oak stands still provide welcome refuge and feeding stops for migratory songbirds before or after crossing the Gulf of Mexico. Grasslands, however, succumbed to the plow in many areas or were invaded by brushy species such as mesquites, acacias, and prickly pears. Large cattle ranches preserve grasslands in some parts of the Coastal Prairie. The region includes the barrier islands and sandy peninsulas that protect bays and lagoons along the coast.

PART TWO

Species Accounts

5 RIVER AND STREAM FAUNA

When you pollute a river, it's a supreme injustice to those who are downstream and those who live in the river who are not human beings. — PAUL HAWKIN

The boundaries of Texas encompass 268,820 square miles—a vast expanse drained by over 3,700 named streams and 15 major rivers (fig. 5.1). Fishes are undoubtedly the most well-known inhabitants of the state's flowing-water ecosystems, but other organisms of remarkable diversity occupy almost every available freshwater habitat. Most aquatic animals are somewhat specialized and live within a relatively narrow range of flow rates, temperatures, and chemical conditions. Many species also require particular substrates for successful reproduction.

Historically, the state contained only the western half of Caddo Lake, its single natural lake, which is shared with Louisiana. To ensure reliable sources of water and to control flooding, humans constructed dams and reservoirs following settlement. Impoundment of water alters the natural flood-pulse cycles of rivers and streams. Riverine ecosystems evolved in response to natural flood-pulse cycles—periodic and somewhat predictable increases and decreases in water levels. Artificial flood-pulse cycles created by sporadic releases from reservoirs significantly modify the conditions under which the native fauna developed and usually cause concomitant decreases in faunal diversity. The water released from impoundments is usually colder than normal streamflow and alters the patterns of sediment deposition. Releases of cold water shock aquatic organisms, reduce their food supply, impair their reproductive cycle, or kill them outright. Increased siltation can suffocate freshwater mussels in their beds and eliminate the habitats required by mussels and fishes, including the fishes serving as hosts for mussel larvae.

Not long after settlers arrived, rivers and streams became convenient places to discard trash and other wastes. For nearly a century, sewage effluents, industrial wastewater, old tires, and other debris fouled the watercourses. While these remain problems in some areas, the main pollutants these days come from urban and agricultural runoff. Most species of river and stream fauna protected as threatened or endangered have experienced population declines resulting from human alterations of water flow or water quality. The following accounts describe the listed river and stream fauna.

Figure 5.1. The lengths, flow rates, and riparian habitats of Texas rivers vary greatly, as exemplified by the salty crusts on the bed of the Little Red River (Hall County) and the clear, deep pool on the Frio River (Uvalde County). Photographs by Leaflet, Creative Commons (left), and Sandra S. Chapman (right).

Freshwater Mussels

NEIL B. FORD AND DAVID F. FORD

Ouachita Rock Pocketbook, *Arkansia wheeleri*
Texas Hornshell, *Popenaias popeii*
Texas Fatmucket, *Lampsilis bracteata*
Texas Fawnsfoot, *Truncilla macrodon*
Texas Pimpleback, *Quadrula petrina*
Smooth Pimpleback, *Quadrula houstonensis*
Golden Orb, *Quadrula aurea*

Freshwater mussels occur worldwide and are among the most endangered fauna on Earth. Like many aquatic species, freshwater mussels are disproportionately imperiled relative to terrestrial species. Numbers of both species and individuals have been declining significantly over the past century. Fifty-two mussel species historically populated Texas waters. Many of these now occupy only small segments of their original ranges, and one species is known to be extinct. An additional species, the False Spike, was thought to be extinct but was recently rediscovered by researchers. Fifteen species are state listed as threatened, one of these is federally listed as endangered, and five are candidates for protection under the Endangered Species Act.

DESCRIPTION. Like clams, oysters, scallops, and other bivalve mollusks, freshwater mussels possess two hard shells (valves) secured together dorsally by a strong, flexible hinge ligament and closed by a pair of robust muscles. The external surfaces of the valves may be smooth, lightly grooved, furrowed, or sculptured with small pimple-like eruptions or larger knob-like bumps. Within a species or population, individual shells may be smooth, while others may exhibit various patterns of bumps or furrows. Shell shape often characterizes a species and, in some cases, a gender. Shells are composed of three layers—two of calcium carbonate compounds and the outer layer (periostracum) of a protein that protects the shell from abrasion and natural acids. In many species, toothlike projections termed "pseudocardinal teeth" (usually two left and one right) mesh to align the two shells. The interior of each valve is lined by a shiny iridescent layer (nacre) deposited by the thin, fleshy mantle, which encloses the living tissues and assists in feeding and respiration.

Bivalves can extend a large muscular foot beyond the valves to maneuver themselves through the substrate. Most of the time, however, freshwater mussels remain immobile, partially buried in the substrate of a lake or stream with the anterior end pointed downward and the posterior end extended just above the sediment surface. Some mussels burrow more deeply into the substrate in response to deceasing water temperatures, increasing current velocity, or disturbances. If lateral movement occurs, it is typically slow and rarely involves great distance.

Mussels are filter feeders, filtering particulates from the water column. With valves slightly open, they draw water into the shell through an incurrent (intake) siphon. The water passes over a pair of gills inside the mantle cavity on each side of the foot. Mucus on the gills captures small food particles—diatoms, algae cells, small organic morsels—and cilia convey them to the mouth. Water exits the mantle cavity via an external siphon carrying wastes released by an anus positioned near the siphon. There is also evidence that mussels are capable of pedal feeding (feeding via their foot).

DISTRIBUTION. Freshwater mussels historically occurred in almost all the state's freshwater habitats, and many rivers once supported extensive mussel populations. Spring headwaters of the rivers of the Edwards Plateau and elsewhere typically support few mussels because springs and rainwater lack adequate organic matter. Rivers subjected to periodic storm-related scouring, especially those below dams, also have fewer mussel species than should be expected.

HABITAT. Each species has its unique physiological tolerances, but some survive in a variety of environments, whereas others occupy a narrow range of habitats. Mussels can inhabit silt, mud, sand, or gravel at the bottom of a stream, river, pond, or lake, but most require stable substrates. Rock slabs, shifting or loose sand, or deep soft silt usually support few mussels. Only a few species survive within the still waters of ponds or lakes, often because oxygen levels diminish at greater depths and water movement is lacking. Mussels fare poorly in waters deficient in oxygen and minerals, especially calcium compounds.

NATURAL HISTORY. Mussels are relatively immobile, slow-growing animals that often remain in one location for their entire life span, which for some may be 100 years or more. Most species have separate sexes, and shell shape in some species varies by gender. During the spawning season, males release sperm, which females take in in through the incurrent siphon. Eggs held on the gills are fertilized and are then retained in brood pouches, where they develop into glochidia (a larval stage). When mature, glochidia are released through various means. The Texas Fatmucket, for example, has a mantle flap resembling a small minnow with an eyespot (fig 5.2), which it uses as a lure. When a predatory fish (usually a sunfish) attacks the lure, the mussel forcibly releases the glochidia. To complete the life cycle, glochidia must attach themselves to the fins or gills of a fish host, where they encyst and live as parasites. When fully formed, a juvenile mussel drops off, burrows into the sediment, and assumes a free-living existence for the remainder of its life. This complex life cycle is easily disrupted, which can result in reproductive failure.

REASONS FOR POPULATION DECLINE AND THREATS. Because they are filter feeders, freshwater mussels are sensitive to fluctuations in water flow, temperature, and deterioration in water quality. Most declines in mussel abundance are associated with detrimental effects from human activities, including reservoir construction, dredging, channelization, agriculture, timber harvest, mining, erosion, and industrial and domestic pollution. Reservoirs alter water temperature and chemistry, which impacts mussel reproduction and survival. Dams often block fish migrations and can result in changes to fish assemblages that mussels depend on to complete their life cycle. Siltation, a major factor downstream of dams, can cover and suffocate mussel beds.

Commercial harvest of freshwater mussels no longer poses the threat in Texas that it once did. Caddo Lake received national notoriety in 1909 when the "Great Caddo Lake Pearl Rush" began. Fortune hunters streamed into the area to harvest pearls from the lake's mussels, but only a few got rich. In 1912, the lake yielded pearls valued at $99,200, equivalent to about $1.9 million today, and hundreds of mussels were opened to find a single pearl. Shell-button industries also existed briefly on some Texas rivers until plastic made such buttons obsolete. Until recently, the Asian artificial pearl industry used shell material from the United States for pearl seeds.

A new threat to native mussels has emerged— invasive species. Although Asian Clams, which arrived in Texas in the 1940s, often live in freshwater mussel communities without obvious negative impacts on native species, the recent (2010) expansion of Zebra Mussel populations into Texas waters poses a significant threat. Zebra Mussels frequently encrust all exposed hard surfaces, including the shells of native mussels. Dense accumulations of Zebra Mussels can eliminate native mussel fauna by interfering with valve closure, increasing the energetic costs of locomotion, and reducing food resources.

RECOVERY AND CONSERVATION. Maintenance of the ecological integrity of the remaining healthy mussel assemblages is essential to their conservation. Dependence on specific fish hosts renders mussels susceptible to population declines associated with elimination of host fish species.

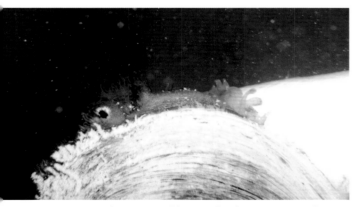

Figure 5.2. The minnow-like mantle of a Texas Fatmucket lures predatory fish near enough to be infected with the mussel's parasitic larvae. Photograph courtesy of the US Fish and Wildlife Service.

Consequently, the conservation of mussels requires not only improving water quality but also managing host fish species, including efforts to guarantee that host fish are available during mussel reproductive periods.

Removal of dams and associated reservoirs might provide the greatest benefit to recovery of dwindling mussel populations. In Texas, however, demands for surface water supplies render dam removal impractical. Therefore, restoration of natural flow conditions downstream of dams may offer the greatest potential for habitat improvement. Modifications to dams to allow releases of warm, oxygen-rich water mimicking spring flood pulses, coupled with the reintroduction of mussels captively propagated or translocated from healthy populations, could be beneficial conservation measures to protect Texas mussels for the future.

Ouachita Rock Pocketbook,
Arkansia wheeleri

FEDERAL STATUS: Endangered
TEXAS STATUS: Endangered (Likely Extirpated)

DISTINGUISHING FEATURES. The Ouachita Rock Pocketbook is almost round, but the valves taper slightly toward one end (fig. 5.3). The posterior (less pointed) end of the shell possesses slight undulations. The beak is higher than the hinge, and its surface sometimes has two to three weakly developed grooves forming double loops, especially when young. The periostracum, which is iridescent when wet, varies from chestnut brown to black. Three triangular pseudocardinal teeth are prominent, but lateral teeth are shallow ridges. The nacre is white to bluish and becomes iridescent posteriorly. The sexes are similar.

The Ouachita Rock Pocketbook is sometimes confused with two other mussel species, the Pimpleback and the Threeridge. Neither of these has an iridescent outer shell. Pimplebacks are lighter brown and usually have greenish rays extending from the beak to the margins of the shell. The undulations on the posterior of the Threeridge are more pronounced.

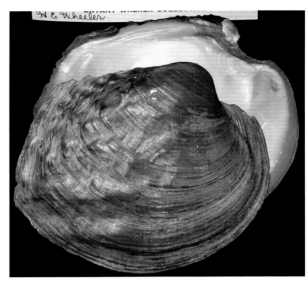

Figure 5.3. The Ouachita Rock Pocketbook. Photograph courtesy of the US Fish and Wildlife Service.

SPECIFIC DISTRIBUTION AND HABITAT. The original specimens of the Ouachita Rock Pocketbook were collected from a series of Ouachita River oxbows north of Arkadelphia, Arkansas. A small population still occurs in the Ouachita River near the type locality. Slightly larger populations inhabit two tributaries of the Red River, the Kiamichi River in southeastern Oklahoma, and the Little River in southeastern Oklahoma and southwestern Arkansas. Single recently dead individuals were recovered from both Pine and Sanders Creeks (Lamar County) at the northern edge of the Texas Piney Woods in the 1990s. No living specimens have been collected from Texas streams or rivers, but these two shells indicate that the species has likely occupied some small northeastern Texas tributaries of the Red River.

The Ouachita Rock Pocketbook tends to inhabit side channels, backwaters, and deep, still pools where large mussel beds contain a diversity of species. Stable gravel or sand substrates adjacent to major riffles and tributary inflows seem to be preferred. The fish hosts of Ouachita Rock Pocketbook have not been confirmed but likely include the Golden Shiner, Duskystripe Shiner, and Freshwater Drum.

FEDERAL DOCUMENTATION:
Listing: 1991. Federal Register 56:54950–54957
Critical Habitat: Not designated
Recovery Plan: US Fish and Wildlife Service. 2004.
Ouachita Rock Pocketbook (*Arkansia wheeleri*
Ortmann and Walker, 1912) recovery plan.
Albuquerque, NM. 83 pp.

Texas Hornshell, *Popenaias popeii*

FEDERAL STATUS: Endangered
TEXAS STATUS: Threatened

DISTINGUISHING FEATURES. The Texas
Hornshell is an elongate, thin, somewhat
trapezoidal mussel that is gently rounded
posteriorly (fig. 5.4). The umbo (the rounded knob
normally forming the highest point on the shell)
barely rises above the hinge and occasionally has
several double-looped or corrugated ridges. The
shell surface, which is olive green, tan, or dark
brown, is mostly smooth, but shallow grooves
may be present dorsally and ventrally, and the
ventral margin exhibits a slight inward curve. The
pseudocardinal teeth (toothlike projections on
the inner edge of the shells) are small, thin, and
flattened above a shallow beak cavity. The nacre is
white and may be tinted with a salmon color near
the beak cavity.

SPECIFIC DISTRIBUTION AND HABITAT. The
Texas Hornshell was historically endemic to the
Rio Grande Basin and many of its drainages in
Texas, Mexico, and New Mexico. The distribution
of the species has declined notably, and it is
currently confirmed to survive within the Rio
Grande and Devils River in Texas and the Black
and Devils Rivers in New Mexico. Three living
individuals were recently found in the lower Pecos
River.

This species prefers areas with large rocks and
boulders but also occupies crevices along Rio
Grande canyon walls and gravel beds below riffles
in the Devils River. It burrows into substrates where
soft sediments collect between boulders, under
rock shelves, and near banks. Such areas also serve
as refuges from flood events. During the breeding
season (January–September), females release
larvae in long, sticky strings of mucus that attach
to the fins, heads, or gills of host fishes. The Texas
Hornshell is a host generalist (capable of using
numerous hosts), but the River Carpsucker, Gray
Redhorse, and Red Shiner are the primary hosts.
Adults may live more than 20 years.

FEDERAL DOCUMENTATION:
Listing: 2018. Federal Register 83:5720–5735
Critical Habitat: Not designated
Recovery Plan: Not developed

Figure 5.4. The Texas Hornshell. Photograph by Joel Lusk, US Fish and Wildlife Service.

Texas Fatmucket, *Lampsilis bracteata*

FEDERAL STATUS: Candidate
TEXAS STATUS: Threatened

DISTINGUISHING FEATURES. The somewhat
oval to elliptical, smooth, greenish- or yellowish-
tan shell of the Texas Fatmucket is decorated
with dark rays extending from the umbo and
broadening toward the ventral margin (fig. 5.5).
The species is sexually dimorphic—males are
more round pointed anteriorly and narrower in
cross section, and females are inflated and bluntly
truncate posteriorly. The umbo barely extends

Figure 5.5. The Texas Fatmucket. Photograph courtesy of the Joseph Britton Freshwater Mussel Collection at the Elm Fork Natural Heritage Museum and the University of North Texas Digital Library.

above the hinge line and bears fine, V-shaped ridges. Internally, the nacre is white tinted with salmon or yellow. The pseudocardinal teeth are narrow, and the lateral teeth curve slightly.

SPECIFIC DISTRIBUTION AND HABITAT. The species is historically endemic to the Guadalupe, San Antonio, San Saba, Llano, Pedernales, Concho, Blanco, and Colorado Rivers and their tributaries, all in central Texas (the Edwards Plateau). Texas Fatmuckets do not occur in portions of these rivers that extend into the Gulf Coastal Plain. This mussel's range has declined significantly, and when found in these rivers, it occurs only in limited numbers.

This mussel occupies firm substrates of mud, sand, or gravel in flowing rivers or tributary streams. Individuals often snuggle into crevices between rock strata or Baldcypress roots. Although the Texas Fatmucket is not known to reproduce in impoundments, some may survive for a limited time near spillways. The minnow-like mantle flap, which acts as a fish lure, can vary considerably in size, shape, and color across the species' range. Known fish hosts (under laboratory conditions) include the Bluegill, Green Sunfish, Guadalupe Bass, and Largemouth Bass.

FEDERAL DOCUMENTATION:
Listing: 2011. Federal Register 76:62166–62212
Critical Habitat: Not designated
Recovery Plan: Not developed

Texas Fawnsfoot, *Truncilla macrodon*

FEDERAL STATUS: Candidate
TEXAS STATUS: Threatened

DISTINGUISHING FEATURES. A Texas Fawnsfoot's relatively thin shell forms an elongated oval that gains thickness toward the anterior end (fig. 5.6). The shell is smooth and has a flattened umbo, which barely rises above the hinge. The shells of males are somewhat more pointed posteriorly and more compressed than those of females. The pseudocardinal teeth are prominent, compressed, triangular, and peg-like and have no major denticles. External color varies individually from tan to shades of brown with reddish or yellowish influences, to dull green, often with faint, interrupted rays and blotches or other markings between the rays. The interior nacre is white to bluish white and is iridescent near the outer edges.

SPECIFIC DISTRIBUTION AND HABITAT. Historically endemic to the Colorado and Brazos drainages of central Texas, the Texas Fawnsfoot was always considered rare, and until recently few individuals were documented. Populations currently survive in the Colorado, Navasota, and San Saba Rivers, as well as the Brazos River and several of its tributaries.

The Texas Fawnsfoot burrows into substrates of mud, sand, or gravel in flowing waters of rivers and larger creeks. It has not been found in impoundments. The fish host for the species is unknown but is likely the Freshwater Drum.

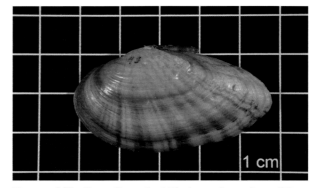

Figure 5.6. The Texas Fawnsfoot. Photograph courtesy of the Joseph Britton Freshwater Mussel Collection at the Elm Fork Natural Heritage Museum and the University of North Texas Digital Library.

FEDERAL DOCUMENTATION:
Listing: 2011. Federal Register 76:62166–62212
Critical Habitat: Not designated
Recovery Plan: Not developed

Texas Pimpleback, *Quadrula petrina*

FEDERAL STATUS: Candidate
TEXAS STATUS: Threatened

DISTINGUISHING FEATURES. The thick shell of the Texas Pimpleback varies considerably in shape among individuals and may reflect the substrate or size of the river in which the animal lives. Shell length is slightly greater than height, giving the mussel a rounded, almost oblong shape (fig. 5.7). A well-developed umbo extends above the hinge and is often adorned with two to four rows of nodules, though these may be absent. The shell surfaces bear parallel ridges, which are more distinct posteriorly. The pseudocardinal teeth are massive, and the lateral teeth are heavy and slightly curved, but the size and shape of these may vary by water body. The interior, which has a deep beak cavity, is white and often iridescent, especially along the posterior margin.

SPECIFIC DISTRIBUTION AND HABITAT. The Texas Pimpleback is endemic to the Colorado and Guadalupe River drainages but no longer occurs in some portions of those river systems. Small populations currently survive in the Concho, Guadalupe, Blanco, San Marcos, Colorado, and San Saba Rivers. The species no longer inhabits the Colorado River upstream of Lake Buchanan.

Mud, sand, and gravel bottoms in rivers and substantial creeks provide habitat for the Texas Pimpleback. Individuals sometimes lodge themselves in gravel-filled cracks in limestone strata. Although the species can be found at depths greater than 6 feet, it does not occur in impoundments. Catfishes are the suspected fish hosts.

FEDERAL DOCUMENTATION:
Listing: 2011. Federal Register 76:62166–62212
Critical Habitat: Not designated
Recovery Plan: Not developed

Smooth Pimpleback, *Quadrula houstonensis*

FEDERAL STATUS: Candidate
TEXAS STATUS: Threatened

DISTINGUISHING FEATURES. The Smooth Pimpleback is a chunky mussel with a shell that ranges from nearly round to a compact, somewhat square oval (fig. 5.8). Its length and height are nearly equal. The mussel is thick in cross section, and the umbo rises well above the hinge but is often heavily eroded. As the name implies, the shells of most specimens are smooth, but some bear a few small pustule-like bumps. The shell may

Figure 5.7. The Texas Pimpleback. Photograph courtesy of the Joseph Britton Freshwater Mussel Collection at the Elm Fork Natural Heritage Museum and the University of North Texas Digital Library.

Figure 5.8. The Smooth Pimpleback. Photograph courtesy of the Joseph Britton Freshwater Mussel Collection at the Elm Fork Natural Heritage Museum and the University of North Texas Digital Library.

be tan, yellowish brown, dark brown, or black, and the nacre is silvery white. The pseudocardinal teeth are fairly heavy, and the right anterior tooth is often chisel shaped. The lateral teeth are short and thick.

SPECIFIC DISTRIBUTION AND HABITAT. The species occurred historically in the Brazos and Colorado River basins of Texas. It currently survives in the San Saba, Leon, Colorado, Navasota, and Brazos Rivers. The species inhabits substrates of mud, sand, or gravel in creeks, rivers, and some reservoirs. Like many other *Quadrula* species, the Smooth Pimpleback utilizes catfishes as its hosts.

FEDERAL DOCUMENTATION:
Listing: 2011. Federal Register 76:62166–62212
Critical Habitat: Not designated
Recovery Plan: Not developed

Golden Orb, *Quadrula aurea*
FEDERAL STATUS: Candidate
TEXAS STATUS: Threatened

DISTINGUISHING FEATURES. Golden or yellowish undertones on an orange, yellowish-brown, or dark brown shell provide the basis for the Golden Orb's name (fig. 5.9). The species tends to be small, with a rectangular, round, or oval form. The umbo rises well above the hinge and bears two to three nodular ridges; the ridges may disappear with age but the nodules sometimes remain. The pseudocardinal teeth are slightly heavy, and the lateral teeth are also pronounced. The nacre is white, sometimes with an iridescent shine.

SPECIFIC DISTRIBUTION AND HABITAT. The Golden Orb once inhabited the Guadalupe, San Antonio, San Marcos, Frio, and Nueces Rivers. Modern occurrences include several locations within the lower San Antonio River, Cibolo Creek, and the Guadalupe, Medina, San Marcos, and Frio Rivers. It also occurs in the Nueces River and one of its reservoirs, Lake Corpus Christi. Many of the extant populations within the San Antonio River are downstream from San Antonio, a rapidly expanding urban center, which could threaten these populations.

Figure 5.9. The Golden Orb. Photograph courtesy of the US Fish and Wildlife Service.

The species occurs primarily within substrates of firm sand, gravel, or mud in the flowing waters of moderately sized rivers and streams. Wave action in Lake Corpus Christi likely simulates the flowing-water conditions required by the species. Although specific host information is lacking, the Golden Orb likely uses Channel Catfish as its host.

FEDERAL DOCUMENTATION:
Listing: 2011. Federal Register 76:62166–62212
Critical Habitat: Not designated
Recovery Plan: Not developed

Fountain Darter, *Etheostoma fonticola*

CODY A. CRAIG
AND TIMOTHY H. BONNER

FEDERAL STATUS: Endangered
TEXAS STATUS: Endangered

The Fountain Darter, endemic to the Guadalupe River basin, was named for standing fountains formed by the voluminous spring discharges from San Marcos Springs before a dam impounded the upper San Marcos River. *Fonticola* means "fountain dweller."

DESCRIPTION. The Fountain Darter is the second-smallest darter in North America (fig. 5.10). The

Figure 5.10. The Fountain Darter. Photograph by Chad Thomas.

total length of the smallest reproductive females is only 0.9 inches, but some attain a maximum length of 2.7 inches. The body is brown or olive dorsally, sometimes with seven to eight dusky dorsal saddles, and white ventrally. Distinct black stitch-like dashes and an incomplete lateral line mark the sides, and the head has dark bars below the eyes. Males display black and red bands on the first dorsal fin, but these colored fin bands are lacking in females.

The Fountain Darter is closely related to the Cypress Darter, but the ranges of the two species do not overlap. The Fountain Darter has naked cheeks and one anal spine, whereas the Cypress Darter has scaled cheeks and two anal spines. Fountain Darters overlap in distribution with several other darter species. These differ from the Fountain Darter by having complete lateral lines and either large dark blotches (Guadalupe Darter) or vertical barring on the sides (Texas Logperch, Greenthroat Darter, and Orangethroat Darter).

DISTRIBUTION. The Fountain Darter is generally restricted to the San Marcos and Comal Rivers and to the spring complexes associated with them. About 103,000 inhabit the upper San Marcos River, where the Fountain Darter represents about 7 percent of the total fish community. Densities are greatest at the headwaters (Spring Lake) and decline near the confluence with the Blanco River, about 5 miles downstream. A single Fountain Darter was found in the San Marcos River about 56 miles downstream from the headwaters (the reference cannot be independently verified).

Approximately 168,000 Fountain Darters inhabit the upper Comal River, where they represent 35 percent of the total fish community. The species

was likely extirpated from the Comal River between the 1950s and 1970s. The current population represents descendants of the 450 fish taken from the San Marcos River and reintroduced to the Comal River in the 1970s.

HABITAT. Fountain Darters are ubiquitous within the spring complexes and rivers but tend to associate more with slack-water and vegetated habitats. The spring complexes provide habitats with stable temperatures (71–74°F) for many spring-associated fishes, but the Fountain Darter survives at 39–90°F and reproduces at 57–79°F— temperatures much cooler and warmer than the normal environments of spring complexes. The Fountain Darter uses aquatic vegetation for feeding, reproduction, and refuge but can conduct these activities in areas lacking aquatic plants.

NATURAL HISTORY. During a protracted spawning season from October through August, a courting male swims near or on top of a female. The female deposits one to three eggs during a spawning event and the eggs are fertilized externally before sinking to the bottom, where they adhere to aquatic vegetation or other surfaces. Female Fountain Darters produce up to 60 eggs per day in multiple batches. Parents provide no care, and the abandoned eggs harden within a few hours of deposition. The larvae hatch four to seven days later. Newly hatched larvae feed on zooplankton and reach sexual maturation within six months. Adult Fountain Darters are sedentary, and when environmental conditions are stable during a one-year period, they move an average of only 33 feet. The adults alter their diet to consume a variety of small invertebrates, including cladocerans, copepods, amphipods, and aquatic insects. Fountain Darters live three years in a hatchery setting, but their life span in the wild is unknown.

REASONS FOR POPULATION DECLINE AND THREATS. The San Marcos River population remained stable within the upper San Marcos River between 1890 and 2010, but it has declined in the lower San Marcos River since 1938. Spring-associated fishes are naturally less abundant downstream from spring complexes, and the decline in Fountain Darters in the lower San Marcos River might be related to several instream dams and weirs that impede natural dispersion. The extirpation of the Comal River population was attributed to an application of the piscicide rotenone during a nonnative fish eradication effort in 1951 as well as low flow for 213 days during a regional drought in 1956. Following the addition of 450 Fountain Darters in the 1970s, the Comal River population was reestablished, and its population and those of numerous other spring-associated fishes stabilized.

The Fountain Darter is currently restricted to 7 miles of stream habitat within two independent rivers that rely on the same groundwater resource (the Edwards Aquifer). Threats to the Fountain Darter are similar to the threats to other species associated with the Edwards Plateau's finite groundwater resources. Groundwater harvest often reduces spring flow and causes corresponding declines in spring-associated fish richness, relative abundance, and density. Urbanization and associated instream modifications to water quality, quantity, and stream morphology (e.g., channelization) are persistent threats to spring-associated fishes. Periodic heavy rainfall results in voluminous discharges from karst spring complexes. Upsurges in spring flow rates increase dissolved oxygen levels and seem to promote resistance and resilience to urbanization among spring fish communities. Introductions of nonnative species, especially trematode parasites, are threats to Fountain Darter survival. Recreational activities, also listed as a threat, are popular in the San Marcos and Comal Rivers, but no link between fish declines and water recreation has been identified.

RECOVERY AND CONSERVATION. Recovery goals for the San Marcos and Comal springs complexes, which include other listed aquatic species, include security for species survival and conservation of the integrity and function of the Edwards Aquifer and spring complexes. The US Fish and Wildlife Service has stated that downlisting of the Fountain Darter is feasible. In 2012, local stakeholders implemented the Edwards Aquifer Habitat Conservation Plan (HCP) to ensure that recovery goals would be met while allowing lawful uses and activities associated with ground and surface waters of the spring complexes. Edwards Aquifer HCP protective measures include water quality monitoring, biological monitoring, ecological modeling, and support of a refugium facility. Surface flows are protected with incentives for water use conservation, voluntary suspension of irrigation, and mandatory reductions in groundwater use during declining aquifer levels.

FEDERAL DOCUMENTATION:
Listing: 1970. Federal Register 35:16047–16048
Critical Habitat: 1980. Federal Register 45:47355–47364
Recovery Plan: US Fish and Wildlife Service. 1996. San Marcos and Comal Springs and associated aquatic ecosystems (revised) recovery plan. Albuquerque, NM. 134 pp.
Habitat Conservation Plan: 2012. Edwards Aquifer Recovery Implementation Program. Habitat conservation plan. 414 pp.

Texas Mosquitofishes

CHAD W. HARGRAVE

Big Bend Gambusia, *Gambusia gaigei*
Pecos Gambusia, *Gambusia nobilis*
San Marcos Gambusia, *Gambusia georgei*
Clear Creek Gambusia, *Gambusia heterochir*

The four species of endangered mosquitofishes in Texas inhabit small, isolated spring-fed wetlands, ponds, and spring runs. Because these fishes live in isolated spring systems, they were likely never widespread throughout the state, but they are usually abundant in habitats that still support existing populations. Loss of spring flows and hybridization with the ubiquitous Western Mosquitofish and Largespring Gambusia are the major threats to the persistence of all endangered mosquitofishes in Texas.

DESCRIPTION. These small fishes (maximum length 1.75 inches) are live bearers (give birth to live young). They have a flattened dorsal surface, upturned mouth, deep body, and rounded dorsal and caudal fins. Most mosquitofishes are olive green and have a silvery belly. Black spotting on their flanks and fins is present to varying degrees. They display strong sexual dimorphism—males are usually smaller than females and possess a modified anal fin (gonopodium) used for copulation. Females are typically larger, and their rounded, somewhat distended belly is marked with a dark posterior abdominal spot around the genital pore. The dark spot enlarges with the development of the brood.

DISTRIBUTION AND HABITAT. The endangered mosquitofishes of Texas occur in isolated spring systems and spring-fed creeks in the Texas Hill Country, the foothills of the Davis Mountains, and near the Rio Grande in the Big Bend region. The local habitats supporting mosquitofishes include wetlands, small streams, and channels created by spring outflows. Stable water temperatures and fine silt or sand substrates characterize these habitats. Most also support an abundant submerged and emergent vegetative community. Mosquitofishes are typically associated with low-flow areas such as stream margins or shallow areas within these larger spring-associated habitats.

NATURAL HISTORY. All mosquitofishes engage in internal fertilization and give birth to relatively large, free-swimming offspring. Male mosquitofishes typically coerce females into copulation by chasing and corralling them. During these activities, they often attempt to swing the gonopodium at the female genital pore. However, some males perform mating displays, and laboratory and field observations indicate that some females choose their mate. A successful male transfers sperm to the genital pore of the female using the gonopodium, which possesses hooks to ensure prolonged contact between the male and female and successful fertilization of the eggs. Females give birth to small broods of large offspring multiple times throughout the year. Mosquitofishes are omnivorous. They consume an array of food items ranging from algae, aquatic insect larvae, aquatic snails, and microcrustaceans to the flying adults of small terrestrial insects that oviposit on the water surface.

REASONS FOR POPULATION DECLINE AND THREATS. The endangered mosquitofishes in Texas are restricted to isolated, thermally stable spring outflows and margins of small spring-fed streams. These species have likely always been relatively rare throughout their Texas range. Groundwater withdrawal, channelization for irrigation, and hybridization with introduced mosquitofishes further threaten extant populations.

RECOVERY AND CONSERVATION. Recovery and conservation focus on preserving existing spring flows, maximizing quality habitat where spring flows exist, and promoting genetic integrity by minimizing potential hybridization with other mosquitofish species such as the Western Mosquitofish and Largespring Gambusia. Near the outflows of San Solomon and Phantom Springs in West Texas (Reeves County), Texas Parks and Wildlife Department biologists have cooperated with local governments, private landowners, and interested citizens to construct three spring wetlands (ciénegas) to increase the availability of habitat previously lost through construction of irrigation canals and a lowered water table. These re-created ciénegas have allowed the local populations of these endangered fishes to increase.

Big Bend Gambusia, *Gambusia gaigei*
FEDERAL STATUS: Endangered
TEXAS STATUS: Endangered

DISTINGUISHING FEATURES. The Big Bend Gambusia has a prominent lateral stripe extending from just behind the head to the base of the tail (fig. 5.11). A dusky teardrop is located beneath the eye, and prominent black crescent-shaped spots are present above the lateral stripe. Its dorsal, anal, and caudal fins are tinted light reddish orange. The species differs from the other endangered mosquitofishes by the absence of dark pigmentation on the chin or below the lateral stripe.

SPECIFIC DISTRIBUTION AND HABITAT. The Big Bend Gambusia is restricted to a limited series of springs near Boquillas Crossing and Rio

Figure 5.11. The Big Bend Gambusia. Photograph courtesy of the US Fish and Wildlife Service.

Figure 5.12. The Pecos Gambusia. Photograph by Chad Thomas.

Grande Village in Big Bend National Park. Within this geographic region, the species is endemic to Boquillas Spring and Graham Ranch Warm Springs. Two other aquatic habitats, Croton Springs on Graham Ranch and Willow Tank, now sustain introduced populations. The springs supporting the Big Bend Gambusia have stable water temperatures, silty and sandy substrates, and abundant emergent (e.g., cattails) and submergent vegetation.

FEDERAL DOCUMENTATION:
Listing: 1967. Federal Register 32:4001
Critical Habitat: Not designated
Recovery Plan: US Fish and Wildlife Service.
    1984. Big Bend Gambusia recovery plan.
    Albuquerque, NM. 29 pp.

Pecos Gambusia, *Gambusia nobilis*
FEDERAL STATUS: Endangered
TEXAS STATUS: Endangered

DISTINGUISHING FEATURES. The Pecos Gambusia has a deep body profile with a dark stripe centered on top of the anterior body from the head to the dorsal fin (fig. 5.12). The fish lacks a prominent lateral stripe, but a threadlike stripe is present in some individuals. With movement, the silver sides flash an iridescent blue or yellow, emphasizing the duller olive tones on the upper half of the body. A black teardrop is present below the eye, and dark coloration surrounds the mouth. A row of small spots anchors the base of the dorsal fin, and several small rounded specks are present on the sides.

SPECIFIC DISTRIBUTION AND HABITAT. The Pecos Gambusia occurs in isolated populations in the Pecos River in Texas and southern New Mexico. In Texas, the Pecos Gambusia also inhabits the headwaters of Phantom Spring Lake (Jeff Davis County), San Solomon Spring, Griffin Spring, and East Sandia Springs (Reeves County), and Diamond Y Draw and Diamond Y Springs, which are owned by The Nature Conservancy (Pecos County). The Pecos Gambusia rejects deeper waters to congregate in shallow spring systems or the margins of spring-fed streams. These habitats have stable water temperatures, high calcium carbonate concentrations, and abundant submerged aquatic vegetation.

FEDERAL DOCUMENTATION:
Listing: 1970. Federal Register 35:16047–16048
Critical Habitat: Not designated
Recovery Plan: US Fish and Wildlife Service. 1982.
    Pecos Gambusia (*Gambusia nobilis*) recovery
    plan. Albuquerque, NM. 41 pp.

San Marcos Gambusia, *Gambusia georgei*
FEDERAL STATUS: Endangered
TEXAS STATUS: Endangered

DISTINGUISHING FEATURES. The San Marcos Gambusia differs from other endangered

Figure 5.13. The San Marcos Gambusia. Photograph by Robert Edwards.

Figure 5.14. The Clear Creek Gambusia. Photograph by Ken Thompson, Fishes of Texas (www.fishesoftexas.org).

mosquitofishes by its distinctly crosshatched sides and the prominent dark stripe on the upper edges of the dorsal and caudal fins (fig. 5.13). Unlike other endemic mosquitofishes, it has lemon-yellow dorsal, anal, and caudal fins.

SPECIFIC DISTRIBUTION AND HABITAT. The last known collection of San Marcos Gambusia in the wild occurred in 1981. Historically, the species was restricted to a spring run below the headsprings of the San Marcos River (Hays County). The spring run was shallow, had a muddy bottom and stable temperatures, and lacked vegetation. The species is likely extinct.

FEDERAL DOCUMENTATION:
Listing: 1980. Federal Register 45:47355–47364
Critical Habitat: 1980. Federal Register 45:47355–47364
Recovery Plan: US Fish and Wildlife Service. 1996. San Marcos/Comal (revised) recovery plan. Albuquerque, NM. 134 pp.

Clear Creek Gambusia,
Gambusia heterochir
FEDERAL STATUS: Endangered
TEXAS STATUS: Endangered

DISTINGUISHING FEATURES. The Clear Creek Gambusia is a large-bodied mosquitofish with a dusky teardrop below the eye (fig. 5.14). The dark abdominal spot characteristic of most

mosquitofish is much larger in this species and extends well beyond the area around the anus. The Clear Creek Gambusia differs from the other endangered *Gambusia* species by a row of faint spots on the middle of the dorsal fin, and terminal dark marks forming distinctive concentric marks on many lateral or dorsal scales.

SPECIFIC DISTRIBUTION AND HABITAT. The Clear Creek Gambusia is restricted to the spring-fed headwaters of Clear Creek, a tributary of the San Saba River (Menard County). Dense mats of Coontail, a submergent flowering plant, flourish in the waters of the creek. The vegetation provides a protective environment for the mosquitofish and habitat for its primary food source, a tiny endemic amphipod (*Hyalella texana*). A dam below the headwaters of the spring prevents upstream invasions of predatory fish and the Western Mosquitofish, which can hybridize with the Clear Creek Gambusia. The dam is critical to the continued existence of the species.

FEDERAL DOCUMENTATION:
Listing: 1967. Federal Register 32:4001
Critical Habitat: Not designated
Recovery Plan: US Fish and Wildlife Service. 1980. Clear Creek Gambusia (*Gambusia heterochir*) recovery plan. Albuquerque, NM. 19 pp.

River and Stream Minnows

ROBERT EDWARDS

Rio Grande Silvery Minnow, *Hybognathus amarus*
Devils River Minnow, *Dionda diaboli*

Minnows in the family Cyprinidae constitute the largest family of fishes worldwide, with more than 3,000 species. Approximately 300 of these species, representing more than 50 genera, occur in North America. Cyprinids also constitute the largest family of fishes in Texas, with more than 57 species. The federal government and Texas list two minnows and three shiners (also minnows) as either endangered or threatened.

DESCRIPTION. All minnows possess a bony connection from the air bladder to the Weberian apparatus (middle ear), which aids in hearing. They are physostomous, meaning they have a duct connecting the esophagus with the air bladder that allows the bladder to fill before emptying into the digestive tract. Their jaws lack teeth, but they grind food with pharyngeal teeth on gill rakers (bony processes in the gill chamber). The number of pharyngeal teeth is used to distinguish species. The largest North American species attains a length of about 6 feet, but the native minnows in Texas are far smaller, ranging between 2 and 4 inches in length.

DISTRIBUTION AND HABITAT. Minnows occur in North America, Europe, Asia, and Africa. They occur throughout freshwater habitats in the United States and usually congregate in large schools. Minnows typically inhabit clear, shallow creeks, streams, and rivers with continuously running water and riffles. Most prefer habitats with sand or gravel bottoms. Some also live along lakeshores.

NATURAL HISTORY. All Texas minnows are relatively short-lived, living on average about one year and at most two to three years. Following heavy rains or snowmelt in the spring and summer, the Rio Grande Silvery Minnow spawns many semibuoyant eggs. After external fertilization by waterborne sperm, the eggs float downstream just under the water's surface (sometimes for more than 100 miles) while the embryos develop. After hatching, the larvae move into backwaters and other slack-water areas where they feed on detritus, algae, and other small foods. As they grow, their food habits expand to include both aquatic and terrestrial invertebrates. For the rest of their lives, they travel upstream, where they spawn as adults, generally only once in their lives.

The Devils River Minnow begins reproductive activities in the winter or early spring and continues through the early summer. Individuals build a small gravel nest about 2 inches in diameter or use inactive or active sunfish nests near spring openings. Males defend territories, and spawning occurs when receptive females visit the nests. Fertilized eggs are deposited near the substrate. Little information is available on embryo and larva development.

REASONS FOR POPULATION DECLINE AND THREATS. Most conservation concerns for endangered and threatened minnows center on the region's limited supply of surface water. While most minnows tolerate many extreme environmental conditions, these species are especially susceptible to certain habitat alterations. Impoundment of rivers for agricultural and municipal water supplies disrupts natural stream flows, alters temperature regimes, and fragments streams, thereby reducing the range of some species. Stream fragmentation is likely responsible for the low genetic diversity of the Rio Grande Silvery Minnow and the Devils River Minnow.

The natural flow of impounded streams is often replaced by controlled reservoir releases. Reservoirs usually prevent downstream flooding, but high-water events are critical for successful reproduction in almost all minnow species except the Devils River Minnow. Groundwater pumping and water diversions, especially in West Texas and the Texas Panhandle, also reduce streamflows and limit surface water supplies. Without adequate streamflow, eggs often settle to the stream bottom and become covered with silt or suffocate from insufficient oxygen, causing extensive mortality among embryos and larvae. Similarly, changes in temperature or declines in constant-temperature spring flows have reduced populations of the Devils River Minnow, which requires a relatively narrow range of temperatures for survival. These threats may be intensified by the interbasin water transfers under consideration to supply water to drier parts of the state.

Many other factors contribute to the decline of minnow populations. Minnows, especially the larvae and young, often die when certain agricultural chemicals or other forms of chemical pollution enter waterways. Introductions of nonnative (exotic) species also reduce native fish communities. Some exotics, such as the armored catfishes in San Felipe Creek, compete for food with the Devils River Minnow—their diets are identical. Introduced stream-bank vegetation, such as Saltcedar and Giant Reed, removes water from streams through transpiration, thus reducing flow.

In most cases, no single factor causes the decline of minnow populations—several factors and their synergistic effects often result in significant reductions in population and distribution. For example, the Rio Grande Silvery Minnow is now found in only a small percentage of its former range. Populations of the Arkansas River Shiner, Sharpnose Shiner, and Smalleye Shiner have experienced similar range compressions.

RECOVERY AND CONSERVATION. Conservation of the Rio Grande Silvery Minnow has been outlined in a recovery plan and multiple recovery-related plans by various agencies. An agreement reached among parties involved in multiple lawsuits established objectives of protecting and restoring Rio Grande ecology while allowing development of a reliable water supply to residents of the Albuquerque region.

Recovery objectives include investigating ways for Rio Grande Silvery Minnows to pass diversion dams, enhancing larval survival by allowing floodplains to flood at critical times, and regulating upstream reservoir releases to mimic natural flow regimes. The recovery plan also prescribes reintroduction of the minnow into a stretch of the Rio Grande flowing through Big Bend National Park. In 2008, biologists introduced more than 431,000 Rio Grande Silvery Minnows into the Rio Grande from stocks cultured at the federal hatchery in Dexter, New Mexico. Biologists representing universities and state and federal agencies continue to monitor the population, which is augmented by the annual release of approximately half a million fish.

Rio Grande Silvery Minnow, *Hybognathus amarus*

FEDERAL STATUS: Endangered
TEXAS STATUS: Endangered

DISTINGUISHING FEATURES. The Rio Grande Silvery Minnow is a small (about 3.5 inches in length), nondescript, round-bodied minnow with a small mouth and relatively small eyes (fig. 5.15). It is light greenish yellow on the dorsal surface and cream to white on the ventral sides. The tips of the dorsal and pectoral fins are rounded. The snout is rounded and when viewed from below appears to overhang the upper lip. Many scales along the upper surfaces of the sides are faintly lined with melanophores (cells containing dark black or brown pigments), giving the impression of a crosshatched or diamond grid pattern.

SPECIFIC DISTRIBUTION AND HABITAT. The Rio Grande Silvery Minnow originally inhabited the Rio Grande from near Española, New Mexico, downstream to the Gulf of Mexico, and the Pecos River from Santa Rosa, New Mexico, downstream to the Rio Grande in Texas. It also occupied the Rio Chama and the Jemez River, tributaries of the Rio Grande in New Mexico, but not the Mexican tributaries of the Rio Grande. Now absent from most of its former range, it currently inhabits the Rio Grande between Cochiti and Elephant Butte Reservoirs in New Mexico. Limited numbers from

Figure 5.15. The Rio Grande Silvery Minnow. Photograph by Robert Edwards.

experimental reintroductions also occupy the Rio Grande in the Big Bend region. The Southwestern Native Aquatic Resources and Recovery Center (formerly the Dexter National Fish Hatchery and Technology Center), the Albuquerque BioPark, and the Uvalde National Fish Hatchery maintain captive populations that provide stock for reintroduction efforts in both New Mexico and Texas.

The species occupies calm-water habitats, often with bottoms of sand or other fine material in areas with continuous water flow. It may be found near shorelines, below debris piles (especially in winter), in backwaters, and in shallow pool areas. It rarely frequents deep runs or habitats with high water velocities.

FEDERAL DOCUMENTATION:
Listing: 1994. Federal Register 59:36988–36995
Critical Habitat: 2003. Federal Register 68:8088–8135
Recovery Plan: US Fish and Wildlife Service. 2010. Rio Grande Silvery Minnow (*Hybognathus amarus*) recovery plan. Albuquerque, NM. 210 pp.

Devils River Minnow, *Dionda diaboli*
FEDERAL STATUS: Threatened
TEXAS STATUS: Threatened

DISTINGUISHING FEATURES. The Devils River Minnow is a relatively small, bicolored minnow that reaches a maximum length of about 3 inches (fig. 5.16). The olive dorsal coloration gives way to silver white below a lateral stripe. The silvery stripe extends from the snout and eye to the base of the caudal fin, which is marked by a prominent triangular spot. Dashes above and below each pore mark the lateral line, and dark pigments outline each dorsal and lateral scale, giving the upper body a crosshatched appearance. Juveniles lack some of these characteristics and may appear silver white with a faint lateral stripe.

SPECIFIC DISTRIBUTION AND HABITAT. The Devils River Minnow was endemic to the Devils River at the western edge of the Edwards Plateau and in San Felipe, Sycamore, Pinto, and

Figure 5.16. The Devils River Minnow. Photograph by Ryan Hagerty, US Fish and Wildlife Service.

Las Moras Creeks in the Chihuahuan Desert. The species disappeared from Las Moras Creek when the springs feeding it dried in 1964 and 1971; the addition of chlorine to the swimming pool below the headsprings caused further loss. Drought likely caused the disappearance of the Sycamore Creek population. The Devils River Minnow remains in the Devils River in southern Texas and the Río San Carlos, Río Sabinas, and Río Salado of Coahuila, Mexico. The species is usually found in flowing, spring-fed waters, often over gravel substrates associated with various types of aquatic vegetation.

FEDERAL DOCUMENTATION:
Listing: 1999. Federal Register 64:56596–56609
Critical Habitat: 2008. Federal Register 73:46988–47026
Recovery Plan: US Fish and Wildlife Service. 2005. Devils River Minnow (*Dionda diaboli*) recovery plan. Albuquerque, NM. 9 pp.

Prairie Stream Shiners

DAVID S. RUPPEL, NICKY M. HAHN, JEREMY D. MAIKOETTER, AND TIMOTHY H. BONNER

Smalleye Shiner, *Notropis buccula*
Arkansas River Shiner, *Notropis girardi*
Sharpnose Shiner, *Notropis oxyrhynchus*

Prairie streams associated with the Brazos and Canadian River drainages of Texas support three federally listed minnows known as "shiners" because scales on their sides reflect silvery iridescence. The Smalleye Shiner and the Sharpnose Shiner occupy the upper Brazos River watershed of Texas. The Arkansas River Shiner occurs in the Canadian River of Texas, and an introduced population exists upstream from Red Bluff Reservoir within the Pecos River drainage.

DESCRIPTION. Prairie stream minnows are members of the genus *Notropis* in the family Cyprinidae. The Smalleye Shiner and Arkansas River Shiner are part of a closely related complex within the subgenus *Notropis*. The Sharpnose Shiner is closely related to species within the *Notropis atherinoides* complex. Common morphological characteristics include small size (< 4 inches in total length), straw to silver color, and small amounts of black pigment.

DISTRIBUTION. The historical distributions of the Smalleye Shiner and Sharpnose Shiner are associated with the Brazos River and Colorado River drainages, although their native or introduced status within the Colorado River drainage is unresolved. The Sharpnose Shiner is also reported in the Wichita River of the Red River drainage, attributed to a natural stream capture of a Brazos River tributary by a Red River tributary. Current distributions of Smalleye Shiners and Sharpnose Shiners include the upper Brazos River upstream from Possum Kingdom Reservoir and the lower Brazos River downstream from Lake Brazos. Shiners are rare in the lower Brazos River.

Historically, the Arkansas River Shiner inhabited the Arkansas River basin, which includes the Canadian River of Texas. It currently occupies the Canadian River basin of New Mexico, Texas, and Oklahoma. Small populations may occur in the Cimarron and Arkansas Rivers of Oklahoma, but the current status of these populations is unknown. A likely introduced population was reported in the Washita River of Oklahoma, and an introduced population exists in the Pecos River of New Mexico.

HABITAT. In Texas, prairie stream minnows are typically associated with run habitats and silt and sand substrates within main stem channels. Adults tend to use deeper and swifter water than juveniles. Chemical and physical environments vary naturally within prairie streams. Prairie stream minnows range widely in their tolerances of salinity, water temperatures, and turbidity.

NATURAL HISTORY. Prairie stream minnows produce multiple batches of offspring over a protracted spawning season. Reproductive season is April through September for Smalleye Shiner and Sharpnose Shiner, and April through August for Arkansas River Shiner. The semibuoyant eggs released into the water column float downstream and hatch within 24 to 48 hours. Larvae of the Arkansas River Shiner continue to drift downstream for two to four days. Juvenile fishes reside in downstream riverine habitats and return upstream as adults for spawning. Reproductive success from egg to mobile juvenile stage likely depends on unobstructed river reaches where eggs and larvae have sufficient distance to drift downstream. Minimum size at sexual maturity is about 1 inch in total length for Arkansas River shiners. Life span is two years. Prairie stream minnows are generalist invertivores, consuming aquatic and terrestrial invertebrates, detritus, plant material, and substrates.

REASONS FOR POPULATION DECLINE AND THREATS. Modifications to prairie streams and reductions of surface flows pose major threats to prairie stream minnows. Modifications such as instream dams that fragment river reaches likely disrupt the reproductive cycle of prairie stream minnows. Reductions of surface flows by surface water diversion and groundwater extraction alter prairie streams and decrease available habitats. Introduced Saltcedar also reduces surface flows, a concern especially during dry periods when streamflows are naturally low. Additional threats previously included the use of prairie stream minnows as commercial bait fishes and currently consist of competition and other concerns with introduced fishes.

RECOVERY AND CONSERVATION. Recovery and conservation efforts among federal, state, and local partnerships address maintenance of water quality and quantity, evaluation of instream barriers, propagation and reintroduction programs, and removal of Saltcedar along riverbanks. As defined by the US Fish and Wildlife Service, approximately 623 miles of the upper Brazos River and its tributaries constitute critical habitat for Smalleye and Sharpnose Shiners. Texas Parks and Wildlife Department biologists removed approximately 1,000 shiners from drying pools in the Brazos River during a drought in 2011. About 700 of the minnows were used to establish a captive-propagation program at Texas Tech University, and nearly 300 were reintroduced to the river in 2012.

Critical habitat for the Arkansas River Shiner includes portions of several rivers in four states, including the Canadian River in Texas. Riparian areas along each watercourse are included in the critical habitat designation.

Smalleye Shiner, *Notropis buccula*

FEDERAL STATUS: Endangered
TEXAS STATUS: Not Listed

DISTINGUISHING FEATURES. The Smalleye Shiner is a dorsally arched and ventrally flattened minnow with a long snout and small eyes. Dark pigments outline the scales dorsally on the straw-colored body, creating a separate black dash at the base of the dorsal fin (fig. 5.17). Adults commonly reach 1.7 inches in total length, but the species has a maximum total length of 3 inches. The Smalleye Shiner is distinguished from other co-occurring minnows by its lack of barbels, short intestine, subterminal mouth, naked nape, and position of the dorsal fin in front of the origin of the pelvic fins.

Figure 5.17. The Smalleye Shiner. Photograph by Clinton and Charles Robertson, Creative Commons.

SPECIFIC DISTRIBUTION AND HABITAT. Smalleye Shiners inhabit the main channel of the upper Brazos River upstream from Possum Kingdom Reservoir as well as the lower Brazos River. However, they have not been collected from the lower Brazos River since the mid-1990s. Occupied habitats are characterized by moderate current velocity, moderate depth (> 1.6 feet), and sandy substrate. Smalleye Shiner is common in the upper Brazos River and represents 12 percent of the fish community.

FEDERAL DOCUMENTATION:

Listing: 2014. Federal Register 79:45273–45286
Critical Habitat: 2014. Federal Register 79:45241–45271
Recovery Plan: US Fish and Wildlife Service. 2015. Recovery outline for Sharpnose Shiner and Smalleye Shiner. Arlington, TX. 16 pp.

Arkansas River Shiner, *Notropis girardi*

FEDERAL STATUS: Threatened
TEXAS STATUS: Threatened

DISTINGUISHING FEATURES. The Arkansas River Shiner is a dorsally arched and ventrally flattened minnow with a small, flattened head, rounded snout, and small eyes (fig. 5.18). Scales are light tan and outlined dorsally with dark pigment, creating a separate black dash at the base of the dorsal fin. Paired black dashes are restricted to the anterior portion of the lateral line, and a black

Figure 5.18. The Arkansas River Shiner. Photograph courtesy of the US Fish and Wildlife Service.

chevron is often present at the base of the caudal fin. Adults commonly reach 2 inches in length, with a maximum length of 3.1 inches. In Texas, the Arkansas River Shiner is distinguished from other co-occurring minnows by its lack of barbels, short intestine, subterminal mouth, and position of the dorsal fin in front of the origin of the pelvic fins.

SPECIFIC DISTRIBUTION AND HABITAT. The Arkansas River Shiner is considered a habitat generalist within the Canadian River of Texas and is found in main channels, side channels, backwaters, and pools. Habitat associations are moderate current velocity, shallow depth (< 1.6 feet), and sandy substrate. Subsurface sand ridges are used by larval and adult fish, likely as foraging areas and velocity refuges. The Arkansas River Shiner represents 22 percent of the fish community within the Canadian River of Texas.

FEDERAL DOCUMENTATION:

Listing: 1998. Federal Register 63:64772–64799

Critical Habitat: 2005. Federal Register 70:59808–59846

Recovery Plan: Not developed

Sharpnose Shiner, *Notropis oxyrhynchus*

FEDERAL STATUS: Endangered

TEXAS STATUS: Not Listed

DISTINGUISHING FEATURES. The Sharpnose Shiner is a compressed minnow with a large, slender head that is accentuated by a conical or sharp snout (fig. 5.19). Scales are predominantly straw colored and outlined with dark pigment dorsally, while the lateral scales have a distinct silver sheen. Maximum size of adults can be up to 3.7 inches in total length. The Sharpnose Shiner is distinguishable from other co-occurring minnows

Figure 5.19. The Sharpnose Shiner. Photograph by Clinton and Charles Robertson, Creative Commons.

by its lack of barbels, short intestine, terminal mouth, and position of the dorsal fin well behind the origin of the pelvic fins.

SPECIFIC DISTRIBUTION AND HABITAT. Sharpnose Shiners can be found in the main channel of the upper Brazos River upstream from Possum Kingdom Reservoir as well as in the lower Brazos River. Habitat associations are moderate current velocity, depth of less than 1.6 feet, and sandy substrate, conditions that are most abundant in the upper Brazos River. The Sharpnose Shiner represents 23 percent of the fish community in the upper Brazos River and less than 0.1 percent of the fish community in the lower Brazos River.

FEDERAL DOCUMENTATION:

Listing: 2014. Federal Register 79:45273–45286

Critical Habitat: 2014. Federal Register 79:45241–45271

Recovery Plan: US Fish and Wildlife Service. 2015. Recovery outline for Sharpnose Shiner and Smalleye Shiner. Arlington, TX. 16 pp.

6 CENTRAL TEXAS CAVE FAUNA

The adaptations of cave organisms to specific conditions have also exposed them to vulnerabilities when the conditions are altered. — DAVID C. CULVER

The Texas landscape is riddled with caves, but by far the greatest concentration occurs in the Edwards Plateau and Balcones Escarpment regions. The porous limestones of the plateau are home to one of the highest numbers of cave-adapted species anywhere on Earth. Isolated habitats that have existed for tens of thousands of years under conditions that differ substantially from those of others often become "hot spots" for speciation—the development of special features resulting in distinct species. Only a few highly adapted species can survive in total blackness deep within a cave system. Long-term existence in the dark wilderness is often accompanied by morphological adaptations that include absence of skin pigment, loss or reduction of eyes, and elongation of appendages (fig. 6.1). The specialized characteristics of these species, called troglobites, limit their dispersal ability. As a result, most troglobites remain confined to a single cave or a few nearby caverns and are extremely vulnerable to extinction.

CAVE FORMATION IN CENTRAL TEXAS. Slightly more than 100 million years ago, much of Texas was submerged beneath a vast sea stretching from the Pacific Ocean to the Gulf of Mexico. Over the next 20 to 30 million years, during the Cretaceous Period, billions of tiny marine organisms flourished in the warm sea. When they died, their calcium carbonate shells drifted to the ocean floor and accumulated in thick layers of calcareous marine sediment. Compressed by the weight of water and subsequent depositions of sediment, the layers hardened into thick sheets of limestone. Flat limestone strata underlie the 23,000-square-mile region known as the Edwards Plateau and are exposed in canyons and deep roadcuts along highways in Central Texas (fig. 6.2).

Although some limestone is dense enough for use in construction, other kinds exhibit considerable variation in hardness. Areas of

Figure 6.1. The blind and unpigmented Cave Crayfish, *Orconectes pellucidus*, from an underground stream in Mammoth Cave exhibits the classic features of a troglobite. Photograph courtesy of the National Park Service.

Figure 6.2. The canyon of the Frio River exposes the flat layers of limestone typical of the Edwards Plateau. Photograph by Brian R. Chapman.

softer limestone are susceptible to fracturing and dissolution by mild acids contained in rainwater. As raindrops fall toward the ground, they absorb carbon dioxide from the atmosphere and become small globules of mild carbonic acid. Over millennia, the acid contained in raindrops slowly dissolves soft areas in exposed limestone and produces pitted surfaces, solution pans, and deep holes. In some areas, rainwater percolating down through cracks in the rock eventually dissolves deeper layers and forms numerous subterranean cavities—sinkholes, caves, and interconnected voids. These honeycombed formations, known as karst, allow rainwater to collect in subsurface pools that gradually enlarge to become extensive underground aquifers.

Some caves in the Edwards Plateau are hypogenic—formed by water ascending from underground systems. A few others (e.g., Caverns of Sonora) were shaped when gases rose through a fault zone and mixed with water in an aquifer. The resulting solution of sulfuric acid dissolved limestone layers and left an enlarged cavern after the water drained off.

The Edwards Plateau in Central Texas harbors one of the largest continuous karst regions in the United States. Some of the karst features continue to act as recharge conduits for aquifers and underground stream systems. Others that once contained water dried out and became air-filled caves as regional uplift and canyon formation lowered the water table. The limestone formations of the Edwards Plateau contain more than 4,000 known caves; undoubtedly many more await discovery (fig. 6.3). The aquifers and caves of the Edwards Plateau have existed for millions of years—long enough to permit extensive speciation in the unique ecosystems. As a result, the plateau is one of the most species-rich areas of the world for karst- and spring-endemic invertebrate, salamander, and fish species.

CAVE HABITATS. Within a cave system, some troglobites occupy mesocaverns—locations inaccessible to larger organisms (and humans) where the humidity and temperature favor their survival. These species leave protected retreats only temporarily to forage in larger passageways.

Other cave-adapted organisms inhabit protective microhabitats in tiny cracks or crevices or under rocks. The hidden and secretive nature of such species presents difficulties to scientists attempting to survey caves to determine species presence, species richness, or population size (fig. 6.4).

The eccentric species found deep within caves never venture forth because they cannot survive beyond their lightless habitat where the temperature is stable and the humidity is high. Their subsistence depends entirely on importations of organic matter from the outside world above. Flooding rains sometimes wash leaf litter and other nutrients into caves, and trogloxenes—animals such as cave crickets, Cave Swallows, and bats that reside in caves and regularly exit to feed—also import many nutrients. Bats are a particularly important component of many Texas caves, where they live in some of the largest populations of any mammal on Earth. The energy in their droppings and carcasses forms the basis for many cave food chains. Few terrestrial animals other than Raccoons venture far into caves, but their excreta in the dark zone nourish a fungus eaten by springtails and other cave invertebrates.

REASONS FOR POPULATION DECLINE AND THREATS. The caves and sinkholes of the Edwards Plateau and Balcones Escarpment support a rich fauna, but many organisms remain undescribed. Many caves throughout the region contain one or more endemic species. Some of these are protected as threatened or endangered, and some are candidate species under review for federal protection. However, the listed or candidate species represent only a small fraction of the endemic and range-restricted cave fauna on the Edwards Plateau; most are found in a few Central Texas counties (Bexar, Travis, and Williamson) where awareness is greatest and conservation efforts have concentrated.

All caves and cave creatures remain vulnerable to human disturbances, and other external forces can disrupt cave ecosystems and eliminate the organisms that depend on them. These perturbations include, but are not limited to, habitat loss through urban expansion, highway construction and other development activities,

Figure 6.3. The entrances to most of the more than 4,000 caves in the state are less obvious than this opening to Natural Bridge Cavern. Photograph by James Sumner, Creative Commons.

Figure 6.4. Complex formations, rocks, and voids hinder surveys of species presence, abundance, and diversity. Photograph by Brian R. Chapman.

cave filling or collapse, mining, quarrying or blasting near caves, alteration of drainage patterns, contamination, and human disturbance and vandalism. Tiny invertebrates inhabiting microhabitats under rocks and other refugia are also susceptible to "take" by crushing or other inadvertent damage during a caving adventure or well-intentioned survey. A recent concern involves the predatory impacts of the Red Imported Fire Ant, a nonnative invasive species that can eliminate cave crickets and other species that provide food to cave organisms.

RECOVERY AND CONSERVATION. The preservation of troglobites is contingent on protection of their unique habitats. Consequently, the recovery plan developed for endangered cave-adapted invertebrates focuses on cave conservation rather than species management. After extensive efforts to locate and explore caves

in Central Texas during the 1980s, biologists identified seven "karst faunal regions" based on cave location, geologic continuity, hydrology, and the distribution of 38 troglobitic species. Four karst faunal regions are in Travis and Williamson Counties, and three are in the northern portion of Bexar County (fig. 6.5). Each karst faunal region consists of several karst faunal areas containing one or more caves supporting listed species. Geologic or hydrologic features separate each cave in a karst faunal area from other caves and constitute barriers to species interchange. Thus, a catastrophic event, such as contamination or flooding, that eliminates the fauna in one cave may not destroy the species in another.

The recovery plan for seven endangered karst invertebrates of Williamson and Travis Counties focuses on downlisting these species from endangered to threatened status. The plan calls for protecting in perpetuity a minimum of three

Figure 6.5. Three of the seven designated karst faunal regions containing caves with threatened and endangered species are in Travis and Williamson Counties near Austin (red), and three are near San Antonio in Bexar County (blue).

karst faunal areas within each karst faunal region. Protection of each cave system within a karst faunal area includes the surface area above the cavern with sufficient expanse to ensure adequate moisture, humidity, air flow, and temperature stability within the subterranean habitat. Efforts to minimize water contamination, ensure the import of nutrients, and eliminate the threat of invasive species, especially the Red Imported Fire Ant, are also major considerations.

Tooth Cave Pseudoscorpion,
Tartarocreagris texana

BRIAN R. CHAPMAN

FEDERAL STATUS: Endangered
TEXAS STATUS: Not Listed

Pseudoscorpions, as their common name implies, look like diminutive scorpions, but they lack a tail and a stinger. Although it is easy to overlook these tiny creatures, over 3,300 species occur worldwide, and they occupy a great diversity of habitats. Although most pseudoscorpions inhabit soil, tree cavities, houses, leaf litter, tree bark, or rock, a few species inhabit caves.

DESCRIPTION. Pseudoscorpions, like all arachnids, have only two major body parts, a cephalothorax and an abdomen, along with four pairs of legs and two additional pairs of appendages equipped with pincers on the anterior end of the body. The largest pair of anterior appendages

(chelicerae) are long and slender and tipped with large clawlike pincers that strongly resemble the pincers on a scorpion; the pincers are equipped with venom glands that subdue prey. The smaller pair of anterior appendages (pedipalps) are close to the mouth and are used in feeding (fig. 6.6). The cephalothorax, chelicerae, and pedipalps are golden brown, whereas the abdomen is light tan. The Tooth Cave Pseudoscorpion lacks eyes and eyespots.

SPECIFIC DISTRIBUTION AND HABITAT. The Tooth Cave Pseudoscorpion exists under rocks in only five caves on the Edwards Plateau. These caves, in Travis and Williamson Counties—Tooth Cave, Amber Cave, Kretschmarr Double Pit Cave, Jester Estates Cave, and Airman's Cave—feature high humidity and stable temperatures.

NATURAL HISTORY. The feeding behavior of the Tooth Cave Pseudoscorpion remains unknown, but it likely preys on microarthropods, primarily mites and springtails, which it locates by smell or touch. Its reproductive biology is also unknown.

FEDERAL DOCUMENTATION:
Listing: 1988. Federal Register 53:36029–36033
Critical Habitat: Not designated
Recovery Plan: US Fish and Wildlife Service. 1994. Recovery plan for endangered karst invertebrates in Travis and Williamson Counties, Texas. Albuquerque, NM. 154 pp.

Cave Harvestmen

PAIGE A. NAJVAR

Cokendolpher Cave Harvestman, *Texella cokendolpheri*
Reddell Cave Harvestman, *Texella reddelli*
Bone Cave Harvestman, *Texella reyesi*

Most people recognize a harvestman when they see one, but they know the distinctive creatures by a different common name—"Daddy Longlegs." The legs of most harvestmen are several times longer than the length of their body. The cavernicolous harvestmen are not to be confused with the

Figure 6.6. This unidentified pseudoscorpion from a Central Texas cave exhibits
the features characteristic of the group. Photograph by Dr. Jean Krejca.

Common Daddy Longlegs, which has long,
threadlike legs and often congregates in large
numbers inside the entrance of many caves.

DESCRIPTION. Harvestmen are often mistaken
for spiders, which they resemble. Both spiders
and harvestmen are arachnids that have the
characteristic eight legs, but they are not closely
related. Harvestmen are members of a distinct
order (Opiliones) and have a single pair of
eyes located in a raised protrusion centered
on their cephalothorax. The cephalothorax,
which represents the fused head and thorax, is
so closely bonded to the abdomen that the two
body regions appear as one oval structure (fig.
6.7). The abdomen in spiders (order Araneae) is
separated from the cephalothorax by a noticeable
constriction, and their three to four pairs of eyes
are usually on the edges of the cephalothorax.

DISTRIBUTION. Cave harvestmen inhabit
many caves on the Edwards Plateau. The greatest
concentration of caves containing the species is in
western Travis and Williamson Counties.

HABITAT. The Cokendolpher Cave Harvestman
and the Bone Cave Harvestman are troglobites
that seek the deep, dark recesses of caves where
the temperature and humidity remain stable. Until
a recent discovery of a Reddell Cave Harvestman
population living in a surface location, biologists
assumed that this species was also a troglobite.
Within cave environments, all three species seek
secluded microhabitats when not foraging.

NATURAL HISTORY. Harvestmen prey on small
invertebrates, such as mites and springtails. Unlike
other arachnids, which regurgitate digestive juices
that dissolve their prey before they consume the

Figure 6.7. Although not a Texas cave species, this specimen of *Hadrobromus grandis* shows the fused cephalothorax and raised eyespot characteristic of all harvestmen. Photograph by Bruce Marlin, Creative Commons.

liquified nutrients, Daddy Longlegs eat their food in chunks. Harvestmen reproduce sexually and lay eggs, but the reproductive behavior and most life history aspects of the cave-adapted species are unknown.

Cokendolpher Cave Harvestman, *Texella cokendolpheri*

FEDERAL STATUS: Endangered
TEXAS STATUS: Endangered

DISTINGUISHING FEATURES. This tiny, pale orange harvestman has a body length of 0.05–0.06 inches and relatively long legs (0.5 inches). The

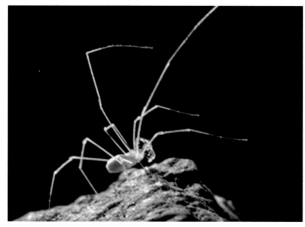

Figure 6.8. The tiny Cokendolpher Cave Harvestman has not been seen in years. It resembles the species shown. Photograph by Dr. Jean Krejca.

second pair of legs are 0.5 inches long (fig. 6.8). The eye mound of the Cokendolpher Cave Harvestman is a narrowed cone adorned with several pointed tubercles.

SPECIFIC DISTRIBUTION AND HABITAT. The Cokendolpher Cave Harvestman is known only from Robber Baron Cave, a large cave in Bexar County. The cave entrance is owned and protected by the Texas Cave Management Association, but urban development surrounds the property and occurs above some reaches of the cavern. Little is known of the harvestman's distribution or habitat within the cave.

FEDERAL DOCUMENTATION:
Listing: 2000. Federal Register 65:81419–81433
Critical Habitat: 2012. Federal Register 77:8450–8523
Recovery Plan: US Fish and Wildlife Service. 2011. Bexar County karst invertebrates recovery plan. Albuquerque, NM. 154 pp.

Reddell Cave Harvestman, *Texella reddelli*

FEDERAL STATUS: Endangered
TEXAS STATUS: Endangered

DISTINGUISHING FEATURES. The 0.08-inch body of an adult Reddell Cave Harvestman is orange, but juveniles are white to yellowish white. The species is distinguished from the other endangered Texas harvestmen by its well-developed eyes on a broadly conical eye mound. Its legs are relatively short—the second pair ranges from 0.2 to 0.3 inches in length.

SPECIFIC DISTRIBUTION AND HABITAT. The species is found in three Jollyville Plateau (Travis and Williamson Counties) caves and four caves in Rollingwood, a suburb of Austin (Travis County). The Colorado River separates the two clusters of caves. Within these caves, the harvestman inhabits lightless zones with a stable temperature and high humidity. It is often collected from the undersides of small rocks on the cave floor. The species was also recently collected at a surface site in Burnet County, indicating that the harvestman is not confined to cave environments.

FEDERAL DOCUMENTATION:

Listing: 1988. Federal Register 53:36029–36033

Critical Habitat: Not designated

Recovery Plan: US Fish and Wildlife Service. 1994. Recovery plan for endangered karst invertebrates in Travis and Williamson Counties, Texas. Albuquerque, NM. 154 pp.

Bone Cave Harvestman, *Texella reyesi*

FEDERAL STATUS: Endangered

TEXAS STATUS: Endangered

DISTINGUISHING FEATURES. The Bone Cave Harvestman was once included as a morphological variant of the Robber Baron Cave Harvestman. It was recognized as a separate species because its eyes are poorly developed or completely absent, it has longer legs (0.2–0.5 inches long), and its orange body is much paler (fig. 6.9). Although it is considered blind, the development of the eye structure varies geographically. All specimens lack retinas, but the eyes of specimens from caves in the southern portion of its range usually possess corneas, whereas those from the northern caves do not. In all localities, the species has a broadly conical eye mound covered with a few small tubercles.

SPECIFIC DISTRIBUTION AND HABITAT. The species inhabits nearly 200 caves in western Travis and Williamson Counties, including Tooth, McDonald, Weldon, Bone, and Root Caves. The Bone Cave Harvestman inhabits the deepest

Figure 6.9. The Bone Cave Harvestman. Photograph by Dr. Jean Krejca.

recesses of caves, where it seeks refuge on the undersides of rocks that lie partially buried in soil.

FEDERAL DOCUMENTATION:

Listing: 1993. Federal Register 58:43818–43820

Critical Habitat: Not designated

Recovery Plan: US Fish and Wildlife Service. 1994. Recovery plan for endangered karst invertebrates in Travis and Williamson Counties, Texas. Albuquerque, NM. 154 pp.

Cave Meshweavers and Spiders

PAIGE A. NAJVAR

Bracken Bat Cave Meshweaver, *Cicurina venii*

Government Canyon Bat Cave Meshweaver, *Cicurina vespera*

Madla Cave Meshweaver, *Cicurina madla*

Robber Baron Cave Meshweaver, *Cicurina baronia*

Government Canyon Bat Cave Spider, *Tayshaneta microps*

Tooth Cave Spider, *Tayshaneta myopica*

Despite extreme variations in body shape among the more than 45,700 spider species found worldwide, most people recognize a spider when they see one. The eight legs, only two body segments, and absence of antennae are distinctive characteristics. Almost all spiders can spin silken webs to capture prey, but the complexity of web design varies considerably. Spiders occur on all continents except Antarctica and dwell in every terrestrial habitat, including freshwater. Many species adapted to existence in caves share characteristics common to most troglobites—loss of functional eyes, reduction of body pigments, and lengthened appendages. More than 70 species occupy limestone caves in Central Texas, but extensive urban development jeopardizes the future of all troglobitic species in the areas surrounding San Antonio and Austin.

DESCRIPTION. The federal list of endangered species contains six species of spiders found in Texas caves. Four are meshweavers (family Dictynidae), which typically construct dense, mesh-like webs on cave floors. Meshweavers are sedentary spiders and have few markings or body features that allow easy identification of species. The other two species are in the family Leptonetidae, a group that generates extensive sheet webs on cave ceilings. The characteristics that distinguish the two spider families are technical and not easily discerned. The six endangered spiders are extremely small and easily overlooked. Females are larger than males, but both sexes use a spinning organ to produce fine silken threads that form a woolly texture when deposited.

DISTRIBUTION. Meshweavers and leptonetid spiders are distributed worldwide, but most of the known species of both groups are found in North America. At least 46 species of eyeless troglobitic meshweavers and about half of the 28 described leptonetid cave spiders occur in Texas caves.

HABITAT. Cave spiders spend most of their lives in dark recesses where the temperature is stable and the humidity is high. They remain near dense sheetlike webs constructed between or under rocks on the cave floor (meshweavers) or on cave walls and ceilings (leptonetid cave spiders).

NATURAL HISTORY. The life history of cave spiders is poorly known, but some information has been gleaned from specimens kept in captivity. Meshweavers live solitary lives in or under dense webs constructed between or under rocks on the cave floor. The woolly mesh webs of some species have a delicate entrance tube with a diameter slightly less than the reach of the spider's extended legs. Some hang upside down in the tube while others station themselves on the cave floor beneath the web. Cave meshweavers devour a variety of small organisms, including isopods, millipedes, springtails, and cave beetles.

Leptonetid cave spiders cover cave walls and ceilings with dense sheet webs. The webs entangle small flying insects and crawling microorganisms. The spiders often dangle from web strands and drop to the cave floor when disturbed.

Like most spiders, cave spiders bite prey with their fangs (chelicerae) and hold them tightly while venom from well-developed glands quiets the victim's struggles. The venom dissolves the

tissues and allows the spiders to ingest the liquified remains. All cave spiders build new webs if their old one is destroyed. Adult males tend to be short-lived because they leave the safety of their abode to search for females. Their searches expose them to predation, and some are killed by females after mating. Epigean (aboveground) species lay eggs in a small silken sac, and troglobitic species likely do the same.

Bracken Bat Cave Meshweaver,
Cicurina venii
FEDERAL STATUS: Endangered
TEXAS STATUS: Not Listed

DISTINGUISHING FEATURES. Only two specimens of the Bracken Bat Cave Meshweaver have ever been collected. The body color of these small, eyeless spider specimens (both are adult females) is cream to light tan (fig. 6.10). The body length is 0.13 inches and the cephalothorax is 0.07 inches long. The species is distinguished from other cave meshweavers by characteristics of the internal anatomy.

SPECIFIC DISTRIBUTION AND HABITAT. The species is known from only two locations in Bexar

Figure 6.10. The Bracken Bat Cave Meshweaver. Photograph by Dr. Jean Krejca.

County. It was initially discovered in Bracken Bat Cave, in the northwestern part of the county. The entrance to the cave is on private property and was filled in 1990. When the species was found at a second location in the path of a planned highway, an overpass was constructed to conserve the habitat. The habitat of the meshweaver is not known.

Government Canyon Bat Cave Meshweaver, *Cicurina vespera*

FEDERAL STATUS: Endangered
TEXAS STATUS: Not Listed

DISTINGUISHING FEATURES. Only one specimen of the Government Canyon Bat Cave Meshweaver has ever been collected. The pale yellow eyeless specimen is only 0.1 inches long (fig. 6.11). The tiny spider's cephalothorax is 0.04 inches long. The species is distinguished from other cave meshweavers by characteristics of the internal anatomy. Molecular analysis indicates that the Government Canyon Bat Cave Meshweaver might be synonymous with the Madla Cave Meshweaver.

SPECIFIC DISTRIBUTION AND HABITAT. The species is known only from Government Canyon

Bat Cave, in a state natural area in northwestern San Antonio. Nothing is known about the distribution of the species within the cave.

Madla Cave Meshweaver, *Cicurina madla*

FEDERAL STATUS: Endangered
TEXAS STATUS: Not Listed

DISTINGUISHING FEATURES. The Madla Cave Meshweaver is a pale tan to cream-colored species lacking eyes (fig. 6.12). Adult females often have a slightly darker stripe running the length of the midabdomen. The average body length is 0.2 inches. The cephalothorax is 0.09 inches long and the abdomen is 0.13 inches long. The species is distinguished from other cave meshweavers by characteristics of the internal anatomy. Recent analysis using molecular methods indicates that the Madla Cave Meshweaver and the Government Canyon Bat Cave Meshweaver are closely related and might be considered conspecific.

SPECIFIC DISTRIBUTION AND HABITAT. The species was described from Madla Cave, about 3 miles north of Helotes in Bexar County. It also inhabits seven other caves in Bexar County. The distribution and habitat occupied by the species in these caves are unknown.

Figure 6.11. The Government Canyon Bat Cave Meshweaver. Photograph by Dr. Jean Krejca.

Figure 6.12. The Madla Cave Meshweaver. Photograph by Dr. Jean Krejca.

Figure 6.13. The Robber Baron Cave Meshweaver. Photograph by Dr. Jean Krejca.

Robber Baron Cave Meshweaver,
Cicurina baronia
FEDERAL STATUS: Endangered
TEXAS STATUS: Not Listed

DISTINGUISHING FEATURES. Adult Robber Baron Cave Meshweavers are eyeless. The body is 0.2 inches long. The cephalothorax, which is 0.1 inches long and 0.07 inches wide, is pale reddish brown, while the cream-colored abdomen is 0.13 inches long (fig. 6.13). The species is distinguished from other cave meshweavers by characteristics of the internal anatomy.

SPECIFIC DISTRIBUTION AND HABITAT. Entry to Robber Baron Cave, the longest known cave in Bexar County, is through a modified sinkhole. Once used as a tourist attraction, the cavern is now protected as a karst resource by the Texas Cave Management Association. The site—Robber Baron Preserve—is surrounded and overlain by a dense urbanized area, but several supervised public visits are conducted each year for educational purposes. The Robber Baron Cave Meshweaver is known only from this cave, but its distribution and habitat in the cavern are unknown.

Government Canyon Bat Cave Spider,
Tayshaneta microps
FEDERAL STATUS: Endangered
TEXAS STATUS: Not Listed

DISTINGUISHING FEATURES. The long-legged Government Canyon Bat Cave Spider is dull yellow and has dusky orange appendages (fig. 6.14). Tiny vestigial structures represent the location of the eyes. Its total length is only 0.07 inches. The first leg is 4.4 times longer than the cephalothorax, which is 0.02 inches in length.

SPECIFIC DISTRIBUTION AND HABITAT. The Government Canyon Bat Cave Spider is known from only two caves in Government Canyon State Natural Area. The natural area protects more than 12,000 acres in northwestern San Antonio from development. Distribution of the species within the caves is not recorded.

FEDERAL DOCUMENTATION (ALL BEXAR COUNTY SPECIES):
Listing: 2000. Federal Register 65:81419–81433
Critical Habitat: 2012. Federal Register 77:8450–8523
Recovery Plan: US Fish and Wildlife Service. 2011. Bexar County karst invertebrates recovery plan. Albuquerque, NM. 154 pp.

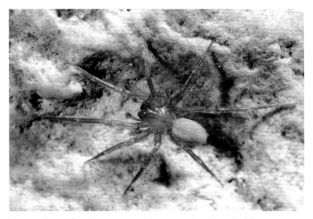

Figure 6.14. The Government Canyon Bat Cave Spider. Photograph by Dr. Jean Krejca.

Figure 6.15. This adult Tooth Cave Spider is about 0.06 inches long. Photograph by Piershendrie, Creative Commons.

Tooth Cave Spider, *Tayshaneta myopica*

FEDERAL STATUS: Endangered
TEXAS STATUS: Not Listed

DISTINGUISHING FEATURES. The body of the whitish Tooth Cave Spider is 0.06 inches long. Its first pair of legs are the longest, measuring a full 0.17 inches, and are 6.1 times longer than longer than the cephalothorax (fig. 6.15). Most specimens are eyeless and depigmented, but some have six eyes and a dull yellow body. The abdomen is whitish to gray.

SPECIFIC DISTRIBUTION AND HABITAT. The species inhabits several caves in Travis and Williamson Counties, including Tooth Cave, Root Cave, Gallifer Cave, and Tight Pit. Its reproductive behavior and life history remain unknown.

FEDERAL DOCUMENTATION:
Listing: 1988. Federal Register 53:36029–36033
Critical Habitat: Not designated
Recovery Plan: US Fish and Wildlife Service.
  1994. Recovery plan for endangered karst
  invertebrates in Travis and Williamson
  Counties, Texas. Albuquerque, NM. 154 pp.

Mold Beetles

PAIGE A. NAJVAR

Kretschmarr Cave Mold Beetle, *Texamaurops reddelli*
Coffin Cave Mold Beetle, *Batrisodes texanus*
Helotes Mold Beetle, *Batrisodes venyivi*

The endangered mold beetles have restricted distributions in caves or near cave entrances on the Edwards Plateau. The caves are imperiled habitats subject to loss or significant environmental change resulting from rapid development occurring on the surface. The troglobitic mold beetles require a stable environment and nutrient inflows that are too easily altered by cave collapse or filling, changes in drainage patterns, modification of surface plant communities, and other environmental modifications.

DESCRIPTION. The mold beetles (order Coleoptera, family Staphylinidae, subfamily Pselaphinae) are often described as ant-like litter beetles. As a group, the small, elongate beetles characteristically possess short wing covers (elytra), which typically do not extend beyond the thorax to cover the first abdominal segment. In the Texas species, the segment forming the end of the antennae is enlarged, and the legs extend well beyond the body. The mold beetles have well-developed mouthparts.

DISTRIBUTION. More than 9,000 species of mold beetles (Pselaphinae) occur worldwide in both surface (epigean) and subterranean habitats. Most of the troglophilic (cave loving) and troglobitic (cave obligate) species inhabit karst regions in the eastern United States and in Missouri, Oklahoma, and Texas. The endangered mold beetles inhabit caves clustered in Bexar, Travis, and Williamson Counties.

HABITAT. Except for the Kretschmarr Cave Mold Beetle, which also occupies surface habitats, the endangered mold beetles are cave dwellers. They occupy the deepest spaces in caves where the temperature is stable and the humidity is high.

Most specimens crawl on the cave floor or hide under rocks buried in silt.

NATURAL HISTORY. Mold beetles in captivity readily eat small organisms. In caves, they likely feed on springtails and mites located by touch or smell. Most dwell beneath rocks buried in soft sediments on the cave floor, but some species find refuge in narrow cracks and crevices. Other aspects of their life history remain unknown.

Kretschmarr Cave Mold Beetle,
Texamaurops reddelli
FEDERAL STATUS: Endangered
TEXAS STATUS: Not Listed

DISTINGUISHING FEATURES. The head has protrusions where the eyes are located on other beetles, but the Kretschmarr Cave Mold Beetle lacks distinct eyes (fig. 6.16). In addition, the posterior abdominal segments are rounded around the body sides and do not present a sharp edge between the top and bottom components. Like other closely related species, the beetle is less than 0.1 inches long and has short wings and long legs.

SPECIFIC DISTRIBUTION AND HABITAT. The Kretschmarr Cave Mold Beetle has been found in four Jollyville Plateau (Travis County) caves but likely occurs in more. It occurs in all cave regions from the entrance and twilight zone to deep, dark recesses. In totally dark cave habitats, it conceals itself under rocks buried in silt. It also occupies

Figure 6.16. The Kretschmarr Cave Mold Beetle. Photograph by Dr. Jean Krejca.

surface habitats where it resides in holes in rotting wood, on the undersurfaces of rocks and logs, in termite nests, and in crevices within the walls and stony debris of sinkholes.

Coffin Cave Mold Beetle,
Batrisodes texanus
FEDERAL STATUS: Endangered
TEXAS STATUS: Not Listed

DISTINGUISHING FEATURES. A groove extending down the middle of the head distinguishes the Coffin Cave Mold Beetle from other closely related mold beetles. The tiny, dark-colored beetle is less than 0.1 inches long. It has reduced eyes, short wings, and long legs. Many characteristics of the Coffin Cave Mold Beetle overlap with those of the Kretschmarr Cave Mold Beetle and *B. cryptotexanus* (no common name), which were once considered a single species that included those now recognized as the Coffin Cave Mold Beetle.

SPECIFIC DISTRIBUTION AND HABITAT. The Coffin Cave Mold Beetle was initially discovered in Inner Space Cavern. Although it inhabits 23 small caves near Georgetown in Williamson County, it does not occur in Coffin Cave. The species has been collected under rocks in clay soil in total darkness where the temperature and humidity are stable.

FEDERAL DOCUMENTATION (KRETSCHMARR CAVE AND COFFIN CAVE MOLD BEETLES):
Listing: 1988. Federal Register 53:36029–36033
Critical Habitat: Not designated
Recovery Plan: US Fish and Wildlife Service.
1994. Recovery plan for endangered karst invertebrates in Travis and Williamson Counties, Texas. Albuquerque, NM. 154 pp.

Figure 6.17. The Helotes Mold Beetle. Photograph by Dr. Jean Krejca.

Helotes Mold Beetle, *Batrisodes venyivi*

FEDERAL STATUS: Endangered
TEXAS STATUS: Not Listed

DISTINGUISHING FEATURES. At first glance, the 0.09-inch Helotes Mold Beetle is easily mistaken for an ant. Its diminutive size, body shape, reddish-brown color, and furtive movements all suggest an ant rather than a beetle (fig. 6.17). The rounded head of this tiny beetle lacks a midline groove. The eyeless beetle possesses short wings and long legs.

SPECIFIC DISTRIBUTION AND HABITAT. Eight caves near Helotes (Bexar County) support populations of the Helotes Mold Beetle. Its distribution and habitat within the cave have not been described.

FEDERAL DOCUMENTATION (HELOTES MOLD BEETLE):
Listing: 2000. Federal Register 65:81419–81433
Critical Habitat: 2012. Federal Register 77:8450–8523
Recovery Plan: US Fish and Wildlife Service. 2011. Bexar County karst invertebrate recovery plan. Albuquerque, NM. 53 pp.

Ground Beetles

WILLIAM B. GODWIN

Ground Beetle (no common name), *Rhadine exilis*
Ground Beetle (no common name), *Rhadine infernalis*
Tooth Cave Ground Beetle, *Rhadine persephone*

Texas ground beetles (*Rhadine* spp.) are notable for their divergence in habitat preferences. Some occupy mountaintop habitats while others are restricted to caves or other subterranean habitats such as mines, cellars, and the tunnels of fossorial (burrowing) species, including gophers or kangaroo rats. Adaptations to subsurface habitats likely served as an important stepping-stone to invading cave systems. Species inhabiting Texas caves may have been restricted to them during warming and drying events in the Pleistocene Epoch and then subsequently adapted to life in darkness.

DESCRIPTION. Ground beetles are elongate, flightless beetles. The cave-adapted ground beetles are slender and have long legs, long antennae, and rudimentary eyes. The eyes of some species are absent. Though pale, the Texas species tend to be reddish brown. Many describe the beetles as "concave" or "depressed" because they usually carry the head and posterior tip of the abdomen higher than the middle of the body.

DISTRIBUTION. Approximately 50 species of ground beetles occur in North America and are distributed from Mexico to Canada. Most species inhabit the southwestern United States, and about 20 occur in the caves of the Edwards Plateau and the Balcones Escarpment. The Colorado River serves as a barrier separating two groups of ground beetles that differ slightly in morphological characteristics.

HABITAT. Troglobitic ground beetles tend to occupy the dark zones of caves, but a few inhabit the twilight zone closer to cave entrances. Most ground beetles dwell where the relative humidity approaches 100 percent. All scamper about on cave floors where the soil is loose or silty.

NATURAL HISTORY. The cavernicolous ground beetles of Edwards Plateau caves are likely predatory generalists. The beetles eat cave cricket eggs and small organisms such as mites, springtails, and immature arthropods. A captive ground beetle accepted and devoured damaged fruit flies, but the flies are probably not a normal part of ground beetle diets. Some ground beetles carry mites attached to their wing covers (elytra) or thorax. The mites are not parasitic—they simply hitch a ride. Other aspects of the ground beetle's life history are unknown.

Ground Beetle (No common name),
Rhadine exilis
FEDERAL STATUS: Endangered
TEXAS STATUS: Not Listed

DISTINGUISHING FEATURES. *Rhadine exilis*, which has not been given a common name, is long and slender. Its head narrows behind the eyespots, which lack the facets found in compound eyes, and the head, pronotum (the most anterior body part), and elytra (wing covers) are much longer than wide (fig. 6.18). The elytra bear striations, and the pronotum has two pairs of marginal setae (tiny hairs).

SPECIFIC DISTRIBUTION. The species inhabits approximately 50 caves in northern and northwestern Bexar County. The beetle rarely inhabits cave entrances.

Ground Beetle (No common name),
Rhadine infernalis
FEDERAL STATUS: Endangered
TEXAS STATUS: Not Listed

DISTINGUISHING FEATURES. This eyeless beetle, *Rhadine infernalis*, is slender and has a pronotum that is 0.7 times as wide as it is long. Its total length is up to 0.4 inches, and its elytra lack striations (fig. 6.19).

SPECIFIC DISTRIBUTION. This beetle is known from only three caves on a single hilltop within the boundaries of Camp Bullis (Bexar County). The US Army, Texas National Guard, and other military organizations use this property for training exercises. Isolated from urban development, the caves receive a degree of protection.

FEDERAL DOCUMENTATION (TWO GROUND BEETLES WITH NO COMMON NAME):
Listing: 2000. Federal Register 65:81419–81433
Critical Habitat: 2012. Federal Register 77:8450–8523
Recovery Plan: US Fish and Wildlife Service. 2011. Bexar County karst invertebrate recovery plan. Albuquerque, NM. 53 pp.

Figure 6.18. *Rhadine exilis*, a ground beetle lacking a common name. Photograph courtesy of the US Fish and Wildlife Service.

Figure 6.19. *Rhadine infernalis*, a ground beetle lacking a common name. Photograph by Dr. Jean Krejca.

Tooth Cave Ground Beetle,
Rhadine persephone
FEDERAL STATUS: Endangered
TEXAS STATUS: Not Listed

DISTINGUISHING FEATURES. Of the several ground beetles that occur in Austin-area caves, the Tooth Cave Ground Beetle is the largest. The reddish-brown adults are up to 0.35 inches long and possess rudimentary eyes. The head is half as wide as long and the antennae extend past the first third of the elytra. The pronotum is short and wide (fig. 6.20).

SPECIFIC DISTRIBUTION. The Tooth Cave Ground Beetle inhabits more than 50 karst caves in Travis and Williamson Counties.

FEDERAL DOCUMENTATION:
Listing: 1988. Federal Register 53:36029–36033
Critical Habitat: Not designated
Recovery Plan: US Fish and Wildlife Service.
    1994. Recovery plan for endangered karst
    invertebrates in Travis and Williamson
    Counties, Texas. Albuquerque, NM. 154 pp.

Figure 6.20. The Tooth Cave Ground Beetle. Photograph by Dr. Jean Krejca.

7 FRESHWATER SPRING FAUNA

When the well goes dry, we know the worth of water.

— BENJAMIN FRANKLIN

Geologists once believed that the crystal-clear water gushing from Central Texas springs owed its origins to melting snow high in the Rocky Mountains. They later learned that the springs stemmed from a complex aquifer system extending from southeastern Oklahoma to West Texas. The northernmost system, the Trinity Aquifer, overlies the Edwards Aquifer for much of its extent, but the two connect hydrologically beneath the Edwards Plateau. The total area of the Trinity-Edwards Aquifer complex encompasses about 77,000 square miles (fig. 7.1).

The Edwards Plateau has few surface streams because rainwater percolates downward through limestone layers perforated like Swiss cheese. Over time, the slightly acidic rainwater etched out cavities where a huge subterranean pool, the Trinity-Edwards Aquifer, collected beneath the plateau. Geologic pressures tilted the plateau's limestone strata slightly from north to south, creating a gradient for water flow though down-dipping interconnected passageways. When rainwater or seepage from sinkholes, creeks, and streams recharges the aquifer, the pressure increases on the water downslope, forcing water to rise through fissures and faults in the overlying strata and to gurgle forth as artesian springs. Other springs emerge where rivers or streams bisect water-bearing limestone layers.

The Trinity-Edwards Aquifer is one of the most prolific artesian aquifers in the world. Most of its approximately 1,900 springs are located along the Balcones Escarpment, which forms the eastern border of the Edwards Plateau and separates it from the Gulf Coastal Plain. The water burbling from springs in Central Texas emerges

Edwards Aquifer

Edwards Aquifer expands across the state of Texas serving nearly two million people and discharging about 900,000 acre feet of water a year. It is also home to many endangered species. "Efforts to improve and conserve the water and springs associated with the Edwards Aquifer will help ensure a healthy future for our community and the plants and animals that depend upon the Edwards Aquifer," said Adam Zerrenner, the Austin Field Office Supervisor.

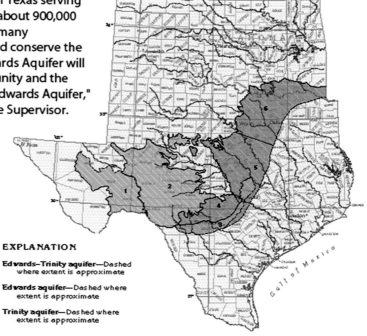

Figure 7.1. The Edwards Aquifer exists beneath a wide swath of Texas and contributes about 900,000 acre-feet of water per year. Many springs along the Balcones Escarpment (zone 3) support endangered species. Map courtesy of the US Geological Survey.

EXPLANATION

Edwards–Trinity aquifer—Dashed where extent is approximate

Edwards aquifer—Dashed where extent is approximate

Trinity aquifer—Dashed where extent is approximate

69

Figure 7.2. Cool, clear springs erupting from the Edwards Plateau limestone often sustain luxuriant microhabitats with a distinct flora and fauna. Photograph by Paige A. Najvar.

Texas Freshwater Amphipods

MARY JONES, DAVID J. BERG, AND NED STRENTH

Diminutive Amphipod, *Gammarus hyalelloides*
Pecos Amphipod, *Gammarus pecos*
Peck's Cave Amphipod, *Stygobromus pecki*

Three described species of federally protected freshwater amphipods occur in Texas, widely separated geographically. Two species are restricted to isolated springs in the Trans-Pecos region, and one is known only from Comal and Hueco Springs in the Guadalupe River watershed of Central Texas.

Many amphipod populations are associated with springs in the northern Chihuahuan Desert in West Texas. These springs feed into the Pecos River and are found along its course from New Mexico to the Rio Grande. Minerals from Permian marine deposits give the water unique chemical properties, including high salinity. The ability to tolerate higher salinity levels may have contributed to ecological speciation in the organisms living in these waters. Recent molecular studies suggest that undescribed species exist in Giffin, San Solomon, and East Sandia Springs (Reeves County) and in Caroline Spring (Terrell County). Once described, these species may also warrant protection as endangered species.

DESCRIPTION. The order Amphipoda is assigned to the class Crustacea in the phylum Arthropoda. Commonly known as amphipods, beach hoppers, and scuds, these crustaceans are found globally in and around marine and freshwater environments. They are generally gray, relatively small, and laterally compressed (fig. 7.3). Unlike several other crustacean groups, they lack a hard carapace. Those that exist only in caves or subterranean waters possess little pigment, and their eyes are reduced or absent.

REASONS FOR POPULATION DECLINE AND THREATS. The primary causes of population decline for these amphipods are habitat alteration and groundwater depletion. Disruptive

at a constant year-round temperature, creating stable environments for aquatic organisms and moist microenvironments in a semiarid region. Consequently, the luxuriant zone around each spring supports a distinct flora and fauna (fig. 7.2). Isolated for thousands of years from related organisms at other spring locations, many organisms adapted so well to local conditions that they became distinct species. Just as isolated caves became "habitat islands" harboring unique faunal representatives, springs of Central Texas also came to support many unique and rare species. The harsh terrestrial environment separating the springs effectively prevents dispersal and promotes genetic isolation. Once a rare species is eliminated from a spring, recolonization is unlikely.

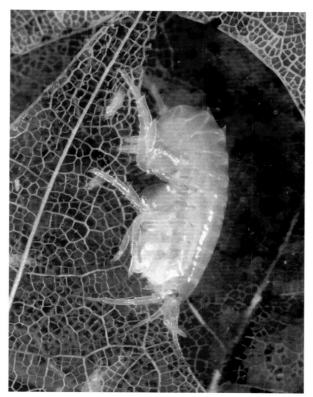

Figure 7.3. All aquatic amphipods share a similar body form. The species depicted here shares many characteristics with the Diminutive Amphipod. Photograph by Scott Bauer, US Department of Agriculture.

human activities, including industrial oil and gas operations and groundwater pumping, are prevalent in the Pecos River valley and throughout the Edwards Aquifer. These factors increase contamination and decrease spring flows, causing subsequent amphipod population declines.

RECOVERY AND CONSERVATION. The US Fish and Wildlife Service has designated critical habitat for these endangered amphipods and for species of aquatic snails from this region. The critical habitat totals 9.2 acres. The habitats of the amphipods are owned and protected by several governmental and private agencies. Habitat for the Diminutive Amphipod is owned by the Texas Parks and Wildlife Department (San Solomon Springs), the US Bureau of Reclamation (Phantom Lake Spring, Jeff Davis County), and The Nature Conservancy (East Sandia Spring). Giffin Springs is privately owned. Access to these areas is closely regulated through each owning entity.

The Diamond Y Spring system in Pecos County is critical habitat for the Pecos Amphipod and two species of aquatic snails (*Pseudotryonia adamantina* and *Tryonia circumstriata*) from this region. The critical habitat totals 441.4 acres and is owned by The Nature Conservancy.

Recovery efforts for the Peck's Cave Amphipod are outlined in the Edwards Aquifer Habitat Conservation Plan. The Edwards Aquifer Recovery Implementation Program (EARIP), established in 2006, is a collaborative process that closely monitors and continues to develop conservation strategies for the endangered species. Several partners including the Edwards Aquifer Authority, the Guadalupe-Blanco River Authority, the cities of New Braunfels, San Marcos, and San Antonio, Texas State University, Texas Parks and Wildlife, and the US Fish and Wildlife Service cooperate in the monitoring and conservation efforts. Additional information may be found at the Edwards Aquifer website: www.eahcp.org.

Diminutive Amphipod,
Gammarus hyalelloides

FEDERAL STATUS: Endangered
TEXAS STATUS: Endangered

DISTINGUISHING FEATURES. Specimens of the Diminutive Amphipod were originally collected in 1967 and were thought to belong to *Hyalella azteca*, a common amphipod found in the region. When Cole described the species in 1976, he recognized this morphological similarity by incorporating part of the generic name, *Hyalella*, in the specific epithet for the new species. The Diminutive Amphipod is the smallest known freshwater gammarid amphipod of North America. Its length ranges from approximately 0.2 to 0.3 inches.

SPECIFIC DISTRIBUTION AND HABITAT. The Diminutive Amphipod was first described from specimens collected near the mouth of Phantom Lake Spring, Toyah Creek (Jeff Davis County). They are often found on Muskgrass, a calcareous aquatic plant typically present at these springs. Their current habitat includes Phantom Lake Spring, San Solomon Spring, East Sandia Spring, and

Giffin Spring. These springs (along with Saragosa Spring and other minor springs) make up the San Solomon Springs system in the Toyah Basin of the Trinity-Edwards Aquifer system. The source of this spring system is primarily groundwater from the Apache Mountains and west of the Delaware Mountains but also includes groundwater from the Davis Mountains during flooding events.

FEDERAL DOCUMENTATION:
Listing: 2013. Federal Register 78:41227–41258
Critical Habitat: 2013. Federal Register 78:40970–40996
Recovery Plan: In development

Figure 7.4. Pigmentation and absence of eyes indicate that Peck's Cave Amphipod is a cave-adapted species. Photograph by Joe N. Fries, San Marcos National Fish Hatchery and Technology Center, US Fish and Wildlife Service.

Pecos Amphipod, *Gammarus pecos*

FEDERAL STATUS: Endangered
TEXAS STATUS: Endangered

DISTINGUISHING FEATURES. The relatively large body of the Pecos Amphipod (approximately 0.5–0.6 inches in length) is banded with red or green. Two distinct features aid in identification of the species: the eyes are elongate and kidney shaped, and the coxal plates on the legs are covered by small bristles.

SPECIFIC DISTRIBUTION AND HABITAT. The only known locality for the Pecos Amphipod is Diamond Y Springs near Fort Stockton (Pecos County). This spring system is in the Delaware Basin and its primary water source is the Rustler Aquifer. The amphipods occupy the undersides of rocks at the spring outlet. Specimens have also been collected from Muskgrass and grasses in Leon Creek, where the spring discharges.

FEDERAL DOCUMENTATION:
Listing: 2013. Federal Register 78:41227–41258
Critical Habitat: 2013. Federal Register 78:40970–40996
Recovery Plan: In development

Peck's Cave Amphipod, *Stygobromus pecki*

FEDERAL STATUS: Endangered
TEXAS STATUS: Endangered

DISTINGUISHING FEATURES. The Peck's Cave Amphipod is a stygobiont, an obligatory subterranean species. Uniquely adapted to its underground habitat, its body is white to yellow because it lacks pigment, as well as eyes. The translucent nature of its exoskeleton allows for the contents of the digestive tract to be faintly visible and can add color (typically varying from orange to tan or dark brown) to the overall appearance (fig. 7.4). Peck's Cave Amphipod, which reaches about 0.5 inches in length, is much larger than the two endangered dryopid beetles found in the same habitat. Like most aquatic amphipods, it looks like a small, laterally compressed shrimp. Both the thorax, which is fused to the head, and the abdomen bear legs (eight pairs in total), which differ in shape and function. Two pairs of antennae extend forward from the head.

SPECIFIC DISTRIBUTION AND HABITAT. Peck's Cave Amphipod resides primarily in dark, subterranean reaches of the Edwards Aquifer, but its subterranean habitat and range are unknown. Samples of water from a well drilled into the Edwards Aquifer at Panther Canyon often contain specimens of the amphipod. The amphipod occasionally emerges from the aquifer at Comal Springs and Hueco Springs, where it finds underwater refuge in crevices between rocks and gravel near spring openings. Hueco Springs, the source of a single specimen, is about 4 miles from Comal Springs. There is no geographic location known as "Peck's Cave." The US Fish and Wildlife Service used "cave amphipod" for all members of genus *Stygobromus* and transliterated the common name for this species from the specific epithet (*pecki*).

Survey results and the visible characteristics of this organism indicate that it likely spends most of its life in the subterranean environment. This amphipod has been found around spring orifices and in the interstices of surrounding rocks. The organism is not well suited for the open stream environment and is likely consumed by fish or birds when in surface waters. Its ability to reenter the subterranean environment is not well known.

Little is known about the natural history of Peck's Cave Amphipods. They likely act as predators of small aquatic organisms, or scavengers that capture bits of organic matter with their mouth appendages. Other aquatic amphipods produce a single brood of eggs and carry them in a brood pouch on the underside of the female's body for one to three weeks. Young amphipods are found in Comal Springs year-round, but much remains to be learned about their reproductive biology.

FEDERAL DOCUMENTATION:

Listing: 1997. Federal Register 62:66295–66304

Critical Habitat: 2007. Federal Register 73:39248–39283

Recovery Plan: 2012. Edwards Aquifer Recovery Implementation Program: Habitat conservation plan. 414 pp.

Comal Springs Aquatic Beetles

JERRY L. COOK

Comal Springs Dryopid Beetle, *Stygoparnus comalensis*

Comal Springs Riffle Beetle, *Heterelmis comalensis*

Two aquatic beetles, the Comal Springs Dryopid Beetle and the Comal Springs Riffle Beetle, exist in springs in or near New Braunfels (Comal County)

Figure 7.5. A complex of springs in Comal and Hays Counties supports numerous endangered invertebrates. Map by Brian R. Chapman.

and San Marcos (Hays County). Both aquatic beetles are listed as endangered because their distributional ranges are limited (fig. 7.5), future spring flows are uncertain, and the potential for contamination of their aquatic habitat is high.

DESCRIPTION. Both the Comal Springs Dryopid Beetle (family Dryopidae) and the Comal Springs Riffle Beetle (family Elmidae) are tiny beetles with elongate bodies. The downward-flexed head appears to be sunken into the prothorax of these drab-colored beetles. Their legs are relatively long and their antennae are slender. Hardened wing covers (elytra) conceal nonfunctional wings. The legs of the pupae are free from the body (exarate).

DISTRIBUTION. Other species in the family represented by the two beetles are found worldwide in many aquatic or terrestrial habitats. Both beetles occur in or near the orifices of Comal Springs and one other major spring. Rising from the Edwards Aquifer, Comal Springs consists of seven major spring apertures and numerous smaller openings

Figure 7.6. Water gushing from outlets of Comal Springs agitates the surface of the Comal River, marking the location where two endangered beetles may occur. Photograph by Larry D. Moore, Creative Commons.

within about 4,300 feet (fig. 7.6). Considered to-gether, the outlets form the largest spring in Texas. The streams emanating from the springs merge to form the headwaters of the Comal River, the shortest river in the United States—it flows for only 2.5 miles before emptying into the Guadalupe River.

San Marcos Springs, from which a single specimen of the Comal Springs Riffle Beetle was taken, flows from over 200 openings along three major limestone fissures. The springs form the headwaters of the San Marcos River, which flows southward 75 miles to its junction with the Guadalupe River. Fern Bank Springs, near Wimberley, supports a small population of the Comal Springs Dryopid Beetle. The Comal Springs Dryopid Beetle also inhabits underground aquatic habitats in the Edwards Aquifer.

HABITAT. The water flowing from the Edwards Aquifer where these species reside maintains a nearly constant temperature between 72° and 74°F. Exiting the limestone aquifer, the waters tend to be neutral or slightly alkaline and normally contain few impurities. The dependability of flow and the stability of thermal and water quality characteristics provided a unique set of ecological

conditions that allowed a number of endemic species to develop. During drought years, however, spring flow declines, sometimes precipitously. These species survived a drought that stopped the flow of Comal Springs from mid-June to early November 1956. During this period, the organisms likely persisted underground in the aquifer or in air-filled cavities within the karst environment.

NATURAL HISTORY. Adult beetles often live for a few years, but they generally reproduce only once per year (a univoltine life cycle). Eggs are deposited on the stems of plants or on the substrate, and the larval cycle lasts for about a year. Dryopid beetles are herbivores as adults and larvae, whereas elmid beetles are either detritivores or algivores. In the aquatic environment, these beetles crawl on stream substrates and do not swim. All species have well-developed tarsal claws that are well adapted to clinging to the stream substrate, emergent vegetation, or stream debris. Unfortunately, the complete life history of these species remains unknown.

REASONS FOR POPULATION DECLINE AND THREATS. These beetles are known only from aquatic habitats associated with the Edwards Aquifer and are completely reliant on water quality and quantity. Growth of the human population in the region has increased the withdrawal of water from the Edwards Aquifer, threatening the permanence and quality of this critical habitat. Urbanization along the corridor between San Marcos and Austin is accompanied by increased probabilities of flooding, siltation, storm water runoff, and pollution, all of which pose threats to the listed species. Of additional concern is the potential for introductions of exotic species that could parasitize or prey on the endangered species, compete with them for food resources, or detrimentally alter the habitat. These species are endemic to the springs in this region and little is known about their biology or resilience to change, but they are thought to be adapted to a stable environment that normally changes little throughout the year. Additionally, the population size of these species is unknown, but the limited size of their habitat suggests that their numbers must be inherently small.

RECOVERY AND CONSERVATION. In 2013, the US Fish and Wildlife Service designated approximately 169 acres as critical habitat to protect the Comal Springs Dryopid Beetle and Comal Springs Riffle Beetle. The critical habitat consists of four units in Comal and Hays Counties, Texas. Because these beetles were recently discovered, there is no information on their historical population levels, nor is there a true sense of their current numbers. The management plan is designed to protect the known habitat of these species, whose numbers are considered low because of their limited range in a habitat under pressure from many external factors. Researchers recently developed marking techniques to monitor populations of the Comal Springs Riffle Beetle, but only limited data are available.

Figure 7.7. An adult Comal Springs Dryopid Beetle taken from Landa Lake, New Braunfels. Photograph by Joe N. Fries, San Marcos National Fish Hatchery and Technology Center, US Fish and Wildlife Service.

Comal Springs Dryopid Beetle,
Stygoparnus comalensis

FEDERAL STATUS: Endangered
TEXAS STATUS: Endangered

DISTINGUISHING FEATURES. The Comal Springs Dryopid Beetle has vestigial (nonfunctional) eyes and a weakly pigmented reddish-brown body that is slightly over 0.12 inches long and 0.04 inches wide (fig. 7.7). The eye remnants consist of several ocellus-like structures. Although the antennae possess eight segments, they are short and not easily observed. The elytra covering the vestigial wings lack prominent punctures or striae (shallow grooves), which are common on many dryopid beetles. A clump of unwettable hairs on their underside holds a thin air bubble when they are submerged, which allows gas exchange so the beetles can respire underwater.

The light yellow or brown larva is cylindrical and elongated, with a partially retracted head.

SPECIFIC DISTRIBUTION AND HABITAT. The Comal Springs Dryopid Beetle is endemic to the Edwards Aquifer and has been collected only in Comal Springs, in both the spring run and on upwellings on the bottom of Landa Lake, and at Fern Bank Springs and from the Panther Canyon well, both in Hays County. Because most specimens have been collected in drift nets, the preferred habitats and habitat requirements are generally unknown. The preferred habitat might be the small areas within the spring that are difficult to sample because of their underground spring association. At one spring upwelling in Landa Lake, specimens were collected on pieces of rotting wood resting on the substrate, but this was likely an incidental association. Larvae have been collected at the base of a hillside near the spring, an expected habitat for dryopid larvae, but too few have been collected to accurately assess larval habitat. Microbial growth and debris associated with this larval habitat are the likely food sources.

Comal Springs Riffle Beetle,
Heterelmis comalensis

FEDERAL STATUS: Endangered
TEXAS STATUS: Endangered

DISTINGUISHING FEATURES. The Comal Springs Riffle Beetle, the smallest member of the family Elmidae in the United States, is only 0.08 inches long and slightly less than 0.04 inches wide (fig. 7.8, left). The body is typical for the genus— elongate with subparallel sides and elytra slightly wider than the prothorax. The surface of the body is clothed with a fine golden pubescence covering

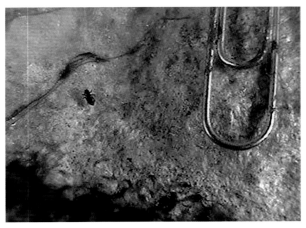

Figure 7.8. The actual size of a Comal Springs Riffle Beetle is evident when compared to a paperclip in the left photo. Photographs by Joe N. Fries, San Marcos National Fish Hatchery and Technology Center, US Fish and Wildlife Service.

the dark brown exoskeleton (fig. 7.8, right). The flight wings are short and nonfunctional. The eyes are well developed, and the threadlike antennae are relatively long. The tibias (middle segment of the legs) have a narrow band of fine hairs on the distal half, and the tarsi (bottom segment of the legs) have both stout and slender spines and well-developed tarsal claws. The spines and claws may help the beetles hold on to and travel across the substrate in moving water. The larvae are elongate and subcylindrical.

SPECIFIC DISTRIBUTION AND HABITAT. The Comal Springs Riffle Beetle is endemic to Comal Springs and San Marcos Springs (Comal and Hays Counties), but it is not a subterranean species. Adult beetles rarely stray more than 8 inches from the spring outlet and seem to prefer areas of low flow where it is dark and around 73°F, and there is an elevated level of dissolved carbon dioxide. Both larvae and adults are aquatic and usually inhabit submerged gravel substrates where aquatic plants are present; some occur where there is no vegetation. Algae and detritus scraped from submerged weeds and rocks appear to provide their food. Pupation likely occurs underwater within the spring.

FEDERAL DOCUMENTATION (BOTH SPECIES):
Listing: 1997. Federal Register 62:66295–66304
Critical Habitat: 2013. Federal Register 78:63100–63127

Recovery Plan: US Fish and Wildlife Service. 1996. San Marcos/Comal (revised) recovery plan. Albuquerque, NM. 93 pp.

West Texas Springsnails

BENJAMIN T. HUTCHINS

Pecos Assiminea, *Assiminea pecos*
Phantom Springsnail, *Pyrgulopsis texana*
Diamond Tryonia, *Pseudotryonia adamantina*
Phantom Tryonia, *Tryonia cheatumi*
Gonzales Tryonia, *Tryonia circumstriata*

The Chihuahuan Desert in West Texas is punctuated by a number of small, isolated springs and seeps issuing from aquifers (underground water systems) deep beneath the parched desert surface. The springs and their associated wetlands provide the only known habitats for several endemic species of fishes, mollusks, crustaceans, and plants. Five species of federally protected aquatic and semiaquatic snails occur in two West Texas spring systems located within a small region. Each species inhabits two or three of the springs and associated habitats, and each co-occurs with one or two of the other listed snails.

DESCRIPTION. Snails are easily recognized members of most terrestrial, marine, and freshwater ecosystems. These mollusks have a muscular foot that extends beyond a single coiled shell to provide movement (fig. 7.9). Most aquatic snails have a hardened, door-like operculum that seals the opening to the shell when the snail withdraws its foot. One or two pairs of tentacles extend from their heads, and the eyes or eyespots of most freshwater snails are located at the base of the first pair of tentacles.

West Texas aquatic snails represent three families of gastropods. The Pecos Assiminea belongs to the family Assimineidae, which includes species that occur in a diversity of marine, freshwater, and terrestrial habitats. The Phantom Springsnail represents the family Hydrobiidae, whereas the Diamond Tryonia, Phantom Tryonia, and Gonzales Tryonia are classified in the closely related family Cochliopidae. Hydrobiids and cochliopids inhabit primarily freshwater and are well represented in the spring systems of western North America. All five of these small species have an operculum and eyespots.

Figure 7.10. West Texas springs in three counties (red) support several endangered species, but some of the underground flow comes from locations farther west (green). Map by Brian R. Chapman.

Figure 7.9. Shells of the five endangered aquatic snails in the Trans-Pecos region are similar to that of the Pecos Assiminea, shown here. Species identifications are based on characteristics of the soft anatomy. Drawing by Sandra S. Chapman.

DISTRIBUTION. The West Texas aquatic snails are confined to two spring systems—San Solomon Spring and Diamond Y Spring—and associated spring runs and wetlands in Jeff Davis, Reeves, and Pecos Counties (fig. 7.10). The spring systems discharge from groundwater flow systems that developed during the Pleistocene Epoch when the climate was much cooler and wetter than today. Modern recharge of the system is limited, but flow rates sometimes increase after periods of heavy rain. The groundwater flow path originates north and west of the springs, extends beneath parts of Hudspeth, Culberson, Jeff Davis, and Reeves Counties, and includes flow between multiple aquifers.

The San Solomon Spring system comprises four spring outflows located close to one another in the Toyah Basin: San Solomon Spring, Giffin Spring, Phantom Lake Spring, and East Sandia Spring. The system has outflows in two counties, Jeff Davis and Reeves. San Solomon Spring, the largest of the four, provides water for irrigation as well as for a large, unchlorinated flow-through swimming pool at Balmorhea State Park. Although the springs were sometimes interconnected after storm events in the past, they are now almost completely isolated by human-made structures.

The Diamond Y Spring system (formerly Willibank Spring) is about 50 miles east of San Solomon Spring, in Pecos County. The spring discharges into a head pool, which drains into a small stream. The outflow stream flows through a broad valley (Diamond Y Draw) where numerous small seeps add volume before the spring run converges with Leon Creek. Several pools and marsh-meadows dot the upper reaches of Leon Creek before the flow dissipates or goes underground. Farther downstream, a smaller spring, Euphrasia Spring, discharges into a head pool that is the source of the lower Leon Creek watercourse. The lower spring run, which is also augmented by small springs and seeps, extends about 0.6 miles.

HABITAT. West Texas aquatic snails inhabit springs and spring runs that discharge directly from karst conduits (in Cretaceous-age limestone) or from fine sediments or gravel in alluvial floodplains, although in the latter case, karst conduits are probably buried under the alluvium. The habitat surrounding the springs and associated wetlands is Chihuahuan Desert grassland or shrubland. Some marshes along watercourses flowing from the springs contain a globally rare plant community. The springs also provide habitat for federally protected fishes, amphipods, and a species of sunflower.

Although regular monitoring of discharge has not been conducted at all sites, most have experienced declines in spring flow since the 1950s. At Phantom Lake Spring, for example, flow rates historically averaged around 10 cubic feet per second (cfs) but have declined to or near zero in recent years. Declines at other sites are not as pronounced. San Solomon Spring, the largest of the five spring complexes, exhibited an average discharge of 30 cfs between 1965 and 2001, which is nearly 10 cfs less than the average reported from the 1930s. Discharge at the remaining springs is less than 5 cfs. The water discharging at all springs is generally clear, well oxygenated, and moderately mineral rich. Groundwater temperatures are typically between 71.6° and 80.6°F, with increasing variability downstream of spring orifices. East Sandia Springs is slightly more saline and cooler (63.5°–73.4°F) relative to the other springs. Most of the springs have undergone substantial human modification, including the draining and degradation of ciénegas (wetland areas characterized by diffuse flow and emergent vegetation), loss of natural habitat in spring runs resulting from the construction of canals for irrigation, and dredging of spring orifices. A large concrete swimming area was constructed over spring orifices of San Solomon Springs. East Sandia Springs and Diamond Y Springs are relatively unmodified (and owned by The Nature Conservancy), but substrates and native vegetation have likely been impacted by livestock grazing and Feral Pigs.

All five snails require permanent, flowing water and become less abundant with increasing distance from spring orifices, probably because of increasing temperature and variability in dissolved oxygen levels. The other four snail species are fully aquatic, associated with flowing water but most commonly found near the margins of channels among emergent vegetation rather than in areas with higher flow velocities. There are some differences in microhabitat preferences.

NATURAL HISTORY. Few natural history studies have been conducted on the federally protected snails occurring in West Texas springs; assumptions about their life history are based primarily on studies of related species. All species are presumed to be grazers of algae and biofilms growing on rocks, mud, and decaying and living vegetation. The snails have a radula (ribbon of rasp-like teeth) that scrapes attached algae and biofilms from substrate surfaces. All species respire through an internal gill.

The species are sexually dimorphic. Reproduction is probably seasonal, but males and females may mate multiple times during the reproductive season. The Pecos Assiminea probably deposits eggs singly or in gelatinous masses attached to a saturated soil surface. The Phantom

Springsnail deposits eggs singly on submerged, hard substrates. The other three species give birth to live young.

The life span of the Pecos Assiminea is unknown, but assimineids in coastal marshes often live 1.5 to 5 years. The snails sometimes reach densities of 65 individuals per square meter in New Mexico. Although the animal is associated primarily with saturated habitats on the periphery of spring runs, it can also be found submerged. The other four species live for one year and may reach densities of more than 10,000 individuals per square meter to more than 100,000 individuals per square meter for some species. Typical densities are much lower and fluctuate from year to year.

REASONS FOR POPULATION DECLINE AND THREATS. Because of their extreme, small-range endemism, many spring-obligate snail species are inherently vulnerable to a diversity of threats. In West Texas, the greatest threat to the long-term persistence of the five federally listed species is declining spring flows. Reduced spring flow and, in some cases, spring failure are well documented in West Texas, including at most sites where the listed species occur. Declining flow is the cumulative effect of drought along with groundwater and surface water extraction for domestic, municipal, agricultural, and oil and gas exploration uses. Declining spring flows may also lead to declines in dissolved oxygen and increases in temperature. Although oxygen and thermal tolerances are unknown for the five species, many aquatic snails have narrow physiological tolerances, making them susceptible to changes in temperature and dissolved oxygen. Biologists have assumed that this is one reason that spring-obligate snails are largely restricted to areas near spring orifices where temperature and dissolved oxygen remain fairly constant. Populations may have once occurred at other, now dry, springs in the area. Spring failure has resulted in the extirpation or extinction of other springsnail species in western North America.

In addition to declining spring flows, many of the springs have been substantially modified, leading in some instances to the degradation or loss of habitat. Modifications include impoundment, dredging, and channelization, primarily for increased use of water for irrigation. These changes probably resulted in extirpation of snails from previously occupied areas within the spring system and may also have reduced population sizes. Although pollution has not been a major threat to the listed snail species to date, several sources of point and non–point source pollution could potentially impact water quality in West Texas springs.

The West Texas aquatic snails also face potential impacts from invasive species. Feral Pigs often frequent the springs where species occur and create wallows that negatively impact water quality, alter flow, and disturb the benthos. Additionally, two species of nonnative aquatic snails, the Red-rimmed Melania and Quilted Melania, also occur at three of the springs. The effects of these nonnative snails on the listed snail species are unknown, but potential negative impacts include predation on snail eggs, disease transfer, and competition for resources.

RECOVERY AND CONSERVATION. The US Fish and Wildlife Service has designated critical habitat to protect springs where the five listed snail species occur. Significantly, however, critical habitat designation does not protect spring flow from the effects of drought or groundwater extraction in the groundwater flow path. Lack of regulatory control over groundwater extraction was one of the major factors cited for the listing of these species. One of the springs is owned and managed by the Texas Parks and Wildlife Department, another by the federal Bureau of Reclamation, two by The Nature Conservancy, and one by a second private landowner.

At San Solomon Springs, two artificial ciénega wetlands were created to restore habitat lost during the channelization and inundation of the springs (fig. 7.11). At Phantom Lake Spring, a pump was installed to supply spring water from inside Phantom Lake Spring Cave to a ciénega wetland area around the spring orifice. The size of the wetland had shrunk substantially, and water flow into the wetland had become intermittent because of declining spring flows. In addition, an artificial ciénega pool was created to connect with the

Figure 7.11. Water from San Solomon Springs near Balmorhea flows into the restored ciénega, which provides habitat for several rare Trans-Pecos aquatic species. Photograph by Larry D. Moore, Creative Commons.

modified wetland around the spring orifice. The creation of the new wetland area was intended to increase available habitat for federally listed snail and fish species, offsetting habitat loss that had resulted from declining spring flows and spring channelization.

Pecos Assiminea, *Assiminea pecos*

FEDERAL STATUS: Endangered
TEXAS STATUS: Endangered

DISTINGUISHING FEATURES. The Pecos Assiminea is a tiny snail bearing a conical shell with a length between 0.06 and 0.07 inches. Although tinted chestnut brown, the shell is nearly transparent. It usually has 4.5 deeply incised whorls and a broad oval opening. Unlike other aquatic snails, the Pecos Assiminea lacks prominent tentacles, and the eyes are located at the ends of short eye stalks.

SPECIFIC DISTRIBUTION AND HABITAT. In Texas, the Pecos Assiminea occupies Diamond Y Spring and East Sandia Spring and their associated drainages. Populations of the species also inhabit two spring systems in Chaves County, New Mexico. The Pecos Assiminea is unique among the five species in that it is rarely found submerged. Rather, it is most commonly associated with saturated,

muddy surfaces and moist vegetation, particularly in and under emergent vegetation near the water's edge along spring runs.

FEDERAL DOCUMENTATION:
Listing: 2005. Federal Register 70:46304–46333
Critical Habitat: 2011. Federal Register 76:33036–33064
Recovery Plan: In preparation

Phantom Springsnail, *Pyrgulopsis texana*

FEDERAL STATUS: Endangered
TEXAS STATUS: Endangered

DISTINGUISHING FEATURES. The tan or dark brown shell of this small snail measures only 0.04 to 0.05 inches long. The shell is described as depressed valvatiform (one to two broad whorls rise above the shell opening, giving the shell a squat appearance). The shell is relatively smooth and has fine growth lines (striations). The thin operculum is yellowish and ovate, and the foot of the snail is dark brown. The species is distinguished from snails of similar size and appearance by mitochondrial DNA analysis.

SPECIFIC DISTRIBUTION AND HABITAT. The Phantom Springsnail is found only in the spring or outflow channels associated with the San Solomon Spring system. It is abundant in the pool at the mouth of Phantom Cave but does not inhabit the cave itself. The snail remains close to spring outflows and high-velocity areas in irrigation canals. Near spring outflows, the Phantom Springsnail occupies both soft and hard substrates and sometimes attaches to submerged plants such as Muskgrass.

Diamond Tryonia, *Pseudotryonia adamantina*

FEDERAL STATUS: Endangered
TEXAS STATUS: Endangered

DISTINGUISHING FEATURES. This small snail has a narrow, tan to dark brown shell that is 0.11 to 0.14 inches long. The shell is relatively smooth and has fine growth lines or striations. The species is distinguished from snails of similar size and appearance by mitochondrial DNA analysis.

SPECIFIC DISTRIBUTION AND HABITAT. The Diamond Tryonia is currently found in Diamond Y Spring, the first mile of the spring run, and seeps with surface water connected to the spring run. It also occurs in the immediate area of the Euphrasia Spring outflow. Within these areas, the snail tends to occupy shallow water near the edges of pools and seeps.

Phantom Tryonia, *Tryonia cheatumi*
FEDERAL STATUS: Endangered
TEXAS STATUS: Endangered

DISTINGUISHING FEATURES. The shell of the Phantom Tryonia is 0.11 to 0.14 inches long. The opaque shell is tan to dark brown and relatively smooth, with fine growth lines or striations. The species is distinguished from snails of similar size and appearance by mitochondrial DNA analysis.

SPECIFIC DISTRIBUTION AND HABITAT. The four springs of the San Solomon Spring system and their associated wetlands are the only places where the Phantom Tryonia is known to occur. The species occupies firm to muddy substrates on the margins of spring outflows and commonly attaches itself to submerged plants, especially dense concentrations of Muskgrass.

Gonzales Tryonia, *Tryonia circumstriata*
FEDERAL STATUS: Endangered
TEXAS STATUS: Endangered

DISTINGUISHING FEATURES. The shell of the Gonzales Tryonia varies in length between 0.11 and 0.14 inches. The tan to dark brown shell surface is marked with well-developed spiral striations. The species is distinguished from snails of similar size and appearance by mitochondrial DNA analysis.

SPECIFIC DISTRIBUTION AND HABITAT. The Gonzales Tryonia occurs only in the outflow

stream of the Diamond Y Spring head pool. Although the species once occupied the head pool associated with Euphrasia Spring in the lower watercourse of the Diamond Y Spring system, it is no longer found there. The snail occupies mud substrates in flowing water on the margins of the springs, seeps, and marshes. It sometimes occupies habitats with emergent cattails and sedges.

FEDERAL DOCUMENTATION (PHANTOM SPRINGSNAIL, DIAMOND TRYONIA, PHANTOM TRYONIA, GONZALES TRYONIA):
Listing: 2013. Federal Register 78:41228–41258
Critical Habitat: 2013. Federal Register 78:40970–40996
Recovery Plan: In preparation

West Texas Pupfishes

ANTHONY A. ECHELLE
AND ALICE F. ECHELLE

Comanche Springs Pupfish, *Cyprinodon elegans*
Leon Springs Pupfish, *Cyprinodon bovinus*

The 50 or so species of *Cyprinodon* are small-bodied, omnivorous egg layers superficially similar to small minnows. Because of their broad environmental tolerances, pupfishes often form dense populations in habitats unsuitable for most other fishes. Their primarily shallow-water, bottom-oriented lifestyle makes them easily observable from shore during most of the year.

The six pupfish species in Texas, with one exception, are subjects of considerable conservation concern. The exception is the wide-ranging Sheepshead Minnow, an abundant coastal species. The remaining species include the Conchos Pupfish in the Río Conchos of Mexico and the Devils River and Alamito Creek in Texas; two genetically distinct forms of Red River Pupfish, one in the Brazos River and the other in the Red, South Canadian, and Colorado Rivers; and three species in the Pecos River system—the federally endangered Comanche Springs Pupfish and Leon Springs Pupfish and the equally imperiled, but not federally listed, Pecos Pupfish.

Figure 7.12. This Pecos Pupfish illustrates the body shape and general characteristics of most pupfishes. Photograph by Chad Thomas, Fishes of Texas (www.fishesoftexas.org).

DESCRIPTION. Pupfishes are generally small (standard length less than 2 inches) and deep bodied, with a short, broad head; a terminal, slightly oblique mouth; a single dorsal fin with no spines; cycloid scales (round, overlapping scales with obvious growth rings); a usually convex or straight terminal edge on the caudal fin; and a single row of spatulate, tricuspid teeth in both jaws (fig. 7.12). The presence of tricuspid teeth distinguishes pupfishes from all other North American fishes. Males typically have deeper, more laterally compressed bodies, larger median fins, and brighter breeding colors than females. Breeding males usually exhibit a conspicuous black band on the terminal edge of the caudal fin and pronounced blue iridescence in the nape region. Juveniles, females, and nonterritorial males are brownish, with dark bars or blotches on the sides; females and juveniles normally have a dark spot or a well-developed black-and-white ocellus (eyespot) at the posterior edge of the dorsal fin.

DISTRIBUTION AND HABITAT. The original geographic ranges of most species do not overlap those of other pupfishes, but some species have been widely introduced. Pupfishes tolerate a wide range of physicochemical conditions and habitats that include relatively stable springs, extremely variable salt-marsh and plains-stream environments, freshwaters of the Florida Lakes region, and hypersaline springs and streams in the American Southwest.

NATURAL HISTORY. Pupfishes are well suited for an opportunistic way of life and for success in harsh habitats. Small body size, omnivory, and broad environmental tolerances allow them to thrive in habitats unsuitable for most other fishes.

Feeding activities are generally bottom oriented, although pupfishes also feed effectively in mid-water and at the surface. Mosquito larvae and other small invertebrates are preferred, but they also ingest algae and bottom substrate materials (e.g., sand and detritus). Consequently, pupfishes have been described as facultative herbivore-detritivores.

Pupfishes enjoy long breeding seasons. Spawning occurs on warm days during most months of the year (all months in some spring-dwelling populations). Breeding males congregate in shallow areas where they form clusters of territories, each defended against all intruders by an intensely aggressive male. When a female prepared to spawn enters the cluster of territories, she is courted by one or a succession of males. With each spawning act, a single egg is released. Afterward, a female may spawn repeatedly with the same male, seek another male, or leave the cluster. Although there is no nest or parental care, the male's territorial activity likely protects eggs from predation. Because sexual maturity is attained two to three months after hatching, some species may produce two or more generations per year.

REASONS FOR POPULATION DECLINE AND THREATS. Most conservation concerns for pupfishes are focused on the American Southwest, where each species is often restricted to a single river or spring system. The restricted distributions of five pupfish species in Texas render them vulnerable to anthropogenic habitat loss as well as competition and hybridization with locally nonnative pupfishes.

Three anthropogenic factors have contributed to most of the loss or degradation of pupfish habitat: (1) damming of rivers to construct large reservoirs, (2) stream channelization and diversion of water flows, and (3) overmining of groundwater (extraction exceeding recharge) for agricultural irrigation or municipal demands. Reservoir construction eliminates shallow-water pools and stream reaches, creates unfavorable physicochemical conditions downstream of the dam, and encourages incompatible biotic communities, which intensifies competition and predation. For example, construction of Amistad, Red Bluff, and Possum Kingdom Reservoirs contributed to declines in, respectively, the Conchos Pupfish, Pecos Pupfish, and Red River Pupfish.

Channelization and diversion of water flows into agricultural fields greatly diminishes the habitat for pupfishes. Off-channel wetlands are reduced or eliminated, and the morphology of the channels precludes development of a thriving population. These effects are particularly severe for the Pecos Pupfish and Comanche Springs Pupfish.

Overpumping of groundwater lowers the water table, ultimately causing springs to dry up. Many springs in Pecos County—including two of the largest in Texas, Comanche Springs and Leon Springs near Fort Stockton—failed for this reason. Prior to the 1940s, Comanche Springs produced about 1 million gallons per hour, and the flow from Leon Springs exceeded 0.5 million gallons per hour. Water from each system flowed downstream before disappearing into the desert (during exceptionally wet periods, their waters likely reached the Pecos River). Spring flows began to decline in the mid-1940s with the advent of heavy groundwater pumping near Belding, southwest of Fort Stockton. By 1962, both spring systems were dry except for occasional flows in Comanche Springs following exceptional rainfall events. The loss of these springs caused the extinction of a morphologically divergent form of Comanche Springs Pupfish and extirpated the Leon Springs Pupfish from much of its native range. Both species persist elsewhere, but the springs they currently occupy are declining, likely because of overmining of groundwater and persistent drought.

West Texas pupfishes are severely threatened by nonnative Sheepshead Minnows, which were introduced into the Pecos River drainage by the 1960s. The Sheepshead Minnow hybridizes with the Comanche Springs Pupfish in a locally restricted situation, and the species was also introduced into the last remaining portion of the Leon Springs Pupfish's native range, which increased the threat of hybrid populations replacing the native species. The Pecos Pupfish was replaced by Sheepshead Minnow × Pecos Pupfish hybrids throughout most of its native range in Texas and New Mexico. Pure

Pecos Pupfish persist only in a 3-mile upstream segment of Salt Creek (Reeves County, Texas), and a small area near Roswell, New Mexico. The genetically distinct Brazos River form of the Red River Pupfish may face a similar threat—Sheepshead Minnows were found in the Brazos River near Possum Kingdom Reservoir in 2011.

Pollution and siltation from oil and gas production and encroachment by Olney's Bulrush threaten some pupfish habitats in West Texas. Dense stands of bulrush eliminate the open bottom habitat required for successful pupfish spawning. Water-hungry Saltcedars and mesquite along streams and pools reduce water levels, especially during droughts.

RECOVERY AND CONSERVATION. Conservation efforts for Texas pupfishes have focused on the two federally endangered species, Comanche Springs Pupfish and Leon Springs Pupfish, and the imperiled Pecos Pupfish. A proposal to list the Pecos Pupfish as endangered was withdrawn after a multiagency conservation agreement was initiated in 1999.

The Pecos Pupfish Conservation Agreement was developed to prevent the spread of Sheepshead Minnow × Pecos Pupfish hybrids and to establish new, genetically pure populations. Private landowners were enlisted to help establish populations in shrimp-farm ponds within the historical range, monitor existing populations, and prevent further expansion of the Sheepshead Minnow and hybrids in the Pecos River system. The state of Texas developed baitfish regulations, constructed or enhanced barriers to dispersal, extirpated local Sheepshead Minnow populations, and conducted genetic monitoring of the remaining genetically pure natural population. Oil and gas operators near pupfish habitats attempt to prevent contamination of sensitive habitats. Bulrush, mesquite, and Saltcedar have been removed from selected areas, and tiles were placed in the Diamond Y Spring head pool to enhance breeding habitat. Captive populations of both species were established at the Southwestern Native Aquatic Resources and Recovery Center (SNARRC) in Dexter, New Mexico, and hybrid Sheepshead Minnow × Leon Springs Pupfish were supplanted with genetically pure

Leon Springs Pupfish reared at SNARRC. Restoration efforts succeeded in effectively restoring the genetic integrity of the species, but suitable habitat in Diamond Y Draw is declining and is much smaller than what was available 30 years ago.

Comanche Springs Pupfish,
Cyprinodon elegans
FEDERAL STATUS: Endangered
TEXAS STATUS: Endangered

DISTINGUISHING FEATURES. Male Comanche Springs Pupfish exhibit uniform black speckling against a silvery white background on the sides, but they lack the blue iridescence seen in most other pupfishes. The dorsal fin is whitish, and the caudal fin of males has a terminal black band. Females have a light brown to gray ground color with a dark lateral band of partially coalesced, somewhat rectangular blotches, a series of blotches above the lateral band, and a few blotches below the lateral band (fig. 7.13).

SPECIFIC DISTRIBUTION AND HABITAT. The Comanche Springs Pupfish originally occurred in two spring systems separated by 55 miles in the Pecos River drainage. By 1956, overpumping of groundwater had desiccated the Comanche Springs complex, causing the extinction of the morphologically divergent local population. The species survives in three large artesian springs (Phantom Lake, San Solomon, and Giffin Springs)

Figure 7.14. San Solomon Ciénega near Balmorhea was created to reproduce "natural" conditions for several rare and endangered species. Photograph by William I. Lutterschmidt.

and small gravity-flow springs in the Toyah Creek drainage near Balmorhea and Toyahvale (Reeves and Jeff Davis Counties). The spring flows, which once nourished large marshy habitats (ciénegas), are now directed into irrigation canals. Three refugia (isolated habitats) now support large pupfish populations in the Balmorhea-Toyahvale area. Two refugia receive flow from San Solomon Spring at Balmorhea State Park. One, Clark Hubbs Ciénega, is a shallow channel 570 feet long that is intersected by a small ciénega, and the other, San Solomon Ciénega, covers 2.5 acres and simulates a natural ciénega (fig. 7.14). An artificial ciénega for the Phantom Lake Spring population receives water pumped from Phantom Cave, the original spring source. After natural outflow ceased in 2000, a submersible pump was installed to restore flow from the cave mouth. The species also occurs in the large swimming pool fed by San Solomon Spring at Balmorhea State Park, in a local system of earthen and concrete irrigation canals, in a short 0.5-mile segment of Toyah Creek, and at the terminus of a 0.5-mile segment of an irrigation flume emptying into Lake Balmorhea. A captive population is maintained at the Uvalde National Fish Hatchery in addition to the one at SNARRC.

The species occupies habitats ranging from the deepest waters at the bottom of the Balmorhea State Park swimming pool (depths of 20 feet) to

Figure 7.13. Male (left) and female Comanche Springs Pupfishes. Photograph by Braz Walker (deceased); gift to Anthony A. Echelle and Alice F. Echelle.

shallow waters in earthen ditches and concrete canals with bottoms covered with debris and vegetation such as Muskgrass. Open patches of bottom substrate are required for spawning. In the irrigation canals, the pupfish prefer areas with low-velocity currents. In contrast to most other species of pupfish, the Comanche Springs Pupfish occurs only in relatively fresh waters (~1–3 ppt total dissolved solids).

FEDERAL DOCUMENTATION:

Listing: 1967. Federal Register 32:4001

Critical Habitat: Not designated

Recovery Plan: US Fish and Wildlife Service.
    1980. Comanche Springs Pupfish (*Cyprinodon elegans*) recovery plan. Albuquerque, NM. 32 pp.

Figure 7.15. Male (left) and female Leon Springs Pupfishes. Photograph by Braz Walker (deceased); gift to Anthony A. Echelle and Alice F. Echelle.

Leon Springs Pupfish, *Cyprinodon bovinus*

FEDERAL STATUS: Endangered

TEXAS STATUS: Endangered

DISTINGUISHING FEATURES. The Leon Springs Pupfish is distinguished from the closely related Pecos Pupfish by having a completely scaled belly. Breeding males exhibit little or no barring and a bluish, somewhat iridescent hue on the sides and nape; the pectoral and pelvic fins are pale yellow, as are the posterior and anterior edges of the dorsal and anal fins and the caudal fin between its base and the terminal black band. Nonbreeding males, females, and juveniles have a broken lateral band of somewhat rectangular, partially coalesced dark blotches that are separated from similar but fainter blotches above the lateral band and smaller, more irregular blotches below; ground colors are whitish below the lateral band and gray brown above (fig. 7.15).

SPECIFIC DISTRIBUTION AND HABITAT. Historically, the Leon Springs Pupfish was endemic to the Diamond Y Draw–Leon Creek system, a small spring-fed system about 9 miles north of Fort Stockton. It presently occurs in only two small spring-fed watercourses separated by about 1.9 miles of dry land in Diamond Y Draw. These include the outflow of the Diamond Y Spring in the upper watercourse, and Mansanto Pool,

the associated Euphrasia Spring, and bulrush-choked portions of the lower watercourse farther downstream. The area is protected by The Nature Conservancy in the Diamond Y Spring Preserve, a 4,099-acre tract protecting one of the largest ciénega systems in West Texas.

The present distribution of the species represents less than 10 percent of its historical range, which included Leon Springs, where the pupfish was first discovered. This spring, located upstream of Diamond Y Spring, was inundated by Lake Leon, an artificial impoundment created for irrigation and fishing in 1918. By 1938, the species was no longer found in the Leon Springs area, although it was anecdotally reported from the Lake Leon outflow in the 1940s.

The highest pupfish densities occur in open, shallow waters of springs, pools, and outflow streams and in areas relatively devoid of bulrush. The water has high concentrations of silica, chlorides, and sulfates and is moderately saline: 3–7 ppt in the upper watercourse and 5–17 ppt in the lower watercourse. A salinity of 41.5 ppt was recorded in a large, desiccating pool (Monsanto Pool) at the head of the downstream watercourse.

FEDERAL DOCUMENTATION:

Listing: 1980. Federal Register 45:54678

Critical Habitat: 1980. Federal Register 45:54678

Recovery Plan: US Fish and Wildlife Service. 1985.
    Leon Springs Pupfish (*Cyprinodon bovinus*) recovery plan. Albuquerque, NM. 30 pp.

Central Texas Salamanders

PAIGE A. NAJVAR

Jollyville Plateau Salamander, *Eurycea tonkawae*
Georgetown Salamander, *Eurycea naufragia*
Salado Salamander, *Eurycea chisholmensis*
Barton Springs Salamander, *Eurycea sosorum*
Austin Blind Salamander, *Eurycea waterlooensis*
San Marcos Salamander, *Eurycea nana*
Texas Blind Salamander, *Eurycea rathbuni*

Seven species of federally protected salamanders occur in aquatic habitats associated with the Edwards Aquifer in Central Texas. To facilitate species-specific accounts, the salamanders are grouped geographically. Species in each group live in proximity and sometimes share the same habitat.

DESCRIPTION. These small aquatic salamanders do not metamorphose into terrestrial forms but remain permanently aquatic and retain well-developed, feathery gills throughout their lives. They have distinct fins on their dorsoventrally flattened tails. Those species that reside in mostly subterranean habitats have some degree of cave-adapted morphology, such as reduced eyes and pale or no coloration.

DISTRIBUTION. The seven species occur in seeps, springs, clear spring-fed rivers, aquifers, and water-filled caverns or cracks along the Balcones Escarpment of the Edwards Plateau in Central Texas. Each species has a limited distribution around a spring, a cluster of springs, or an aquifer.

HABITAT. Water in the habitats occupied by the seven species is typically described as "high quality," with low levels of potential contaminants, high levels of oxygen (> 5.5 parts per million), a narrow temperature range (69.8°–71.7°F) that rarely varies, and neutral to slightly alkaline pH levels (pH 7–7.2). Some species move between surface and subsurface habitats, where they seek interstitial spaces (empty voids between rocks) as foraging habitat and protection from predators.

Aquatic plants, leaf litter, and woody debris provide refuges for several species that live in pools or streams.

NATURAL HISTORY. The life cycle of most salamanders involves metamorphosis from aquatic, gill-breathing larvae into lung-breathing, terrestrial adults. However, all the threatened and endangered Central Texas salamanders are neotenic (they do not transform into a terrestrial form). Instead, these aquatic salamanders mature sexually while retaining the characteristic features of larvae, with external gills and a flattened tail. They do not have lungs but breathe primarily through their gills and skin.

Most of what is known about reproduction in these salamander species has been derived from research and observations on the San Marcos and Barton Springs Salamanders in captivity. Observations of captive courtship pairs of Barton Springs Salamanders indicate their courtship behavior is consistent with the tail-straddling walk unique to salamanders within the family Plethodontidae. During courtship, the male deposits a spermatophore that the female picks up before internal fertilization. Female Barton Springs Salamanders deposit clutches of 5 to 39 eggs, while clutches of captive San Marcos Salamanders have had as many as 73 eggs. Hatching of eggs in captivity occurred within 16 to 39 days for Barton Springs Salamanders and 16 to 24 days for San Marcos Salamanders. Because eggs are rarely found on the surface in the wild, these salamanders likely deposit their eggs underground or attach them to plants or rocks.

The detection of juveniles in all seasons suggests that reproduction occurs year-round. However, increases in the abundance of juvenile Georgetown Salamanders in spring and summer indicate that relatively more reproduction may occur in winter and early spring in this species compared to other seasons.

Unlike terrestrial salamanders, which usually protrude their tongue to feed, these species part their jaws rapidly to create a suction that draws in food. The diet consists of small aquatic invertebrates such as amphipods, copepods, insect larvae, and snails. Known predators of these salamander species in surface habitats include centrarchid fishes (carnivorous fishes belonging to the sunfish family), crayfish, and large aquatic insects.

As their names suggest, the Austin Blind and Texas Blind Salamanders lack functional eyes and spend much of their lives in complete darkness. They are adapted to their subterranean habitats by their use of smell and possibly by their ability to detect vibrations to identify prey and potential mates. Because their habitats are not accessible to biologists, little is known about these two species.

REASONS FOR POPULATION DECLINE AND THREATS. The future of the Central Texas salamanders depends on continuous flows of high-quality spring water. Expansive urban development in the San Antonio and Austin areas has increased dependence on water drawn from the Edwards Aquifer and has amplified contaminant levels in regional seeps, springs, and streams. Demands for groundwater on which these species depend, drought, and the expansion of impervious cover in recharge zones has lowered the water table in the Edwards Aquifer and reduced spring flows. Diminished spring flows for extended periods often result in low dissolved oxygen levels detrimental to the salamanders.

Urbanization also increases levels of contaminants and pollutants in aquatic ecosystems generated by construction activities, wastewater discharge, transportation infrastructure, storm water runoff, and hazardous materials spills. Such impurities affect respiration, growth, reproduction, and survival of aquatic organisms. Surface habitat modifications caused by artificial impoundments, Feral Pigs, livestock, and other human activities often increase the frequency of flooding, which carries silt and debris that uproots aquatic plants and clogs interstitial habitat spaces with sediment. The limited ranges of these species render them especially vulnerable to potentially catastrophic water quality changes and contaminants. Additional endemic *Eurycea* species on the Edwards Plateau likely face similar threats.

RECOVERY AND CONSERVATION. Conservation of these salamanders and their habitats depends on the maintenance of aquifer levels and spring flows. Twenty years after the Edwards Aquifer Authority was created to regulate groundwater withdrawal, stakeholder organizations collaborated to develop an Edwards Aquifer Habitat Conservation Plan to support the recovery of federally listed species that rely on the aquifer. This plan regulated aquifer groundwater pumping and implemented measures to ensure continuous spring flows. The Texas Commission on Environmental Quality and the US Fish and Wildlife Service developed voluntary water quality protection measures to minimize the effects of urban development on the aquatic habitats within the Edwards Aquifer region. These voluntary measures supplement the Edwards Rules, which are state regulations that require water quality protection measures for new developments located over the Edwards Aquifer.

The city of Austin enacted water quality ordinances to protect salamander habitats within Travis County. The city also manages the Barton Springs complex as a recreation area under a federal permit and associated habitat conservation plan, which outline conservation measures designed to protect Barton Springs and Austin Blind Salamanders (fig. 7.16). City of Austin biologists conduct regular surveys of salamander habitat and populations, fund and conduct salamander research and education programs, and maintain a salamander refugium and captive-breeding program for both species. In addition, the city of Austin's Water Quality Protection Lands program protects over 28,000 acres through conservation easements within the recharge and contributing zones of the Barton Springs segment of the Edwards Aquifer. The ecological restoration and protection of these lands help improve water quality and aquifer recharge.

The US Fish and Wildlife Service's San Marcos Aquatic Resources Center maintains a refugium and captive-breeding program for many of the federally listed species in the region, including the San Marcos and Texas Blind Salamanders. Staff at this facility and at Texas State University's Edwards

Figure 7.16.
Barton Springs
serves the city
of Austin as
a recreation
area while
maintaining
essential
habitat for the
Barton Springs
Salamander and
the Austin Blind
Salamander.
Photograph by
Paige A. Najvar.

Aquifer Research and Data Center also conducts ongoing research and monitoring for these species and the Edwards Aquifer.

The Williamson County Conservation Foundation works to conserve federally listed species in Williamson County, Texas. This foundation has purchased lands known to protect habitat for the Georgetown Salamander and other federally listed species. It also funds monitoring and research to further the conservation of this species. In 2016, The Nature Conservancy purchased a conservation easement over 256 acres of the Solana Ranch in Bell County, Texas, with funding acquired through an Endangered Species Act grant. The easement will permanently protect three spring outlets occupied by Salado Salamanders from urban development.

Northern Segment of the Edwards Aquifer

Jollyville Plateau Salamander, *Eurycea tonkawae*
FEDERAL STATUS: Threatened
TEXAS STATUS: None

Georgetown Salamander, *Eurycea naufragia*
FEDERAL STATUS: Threatened
TEXAS STATUS: None

Salado Salamander, *Eurycea chisholmensis*
FEDERAL STATUS: Threatened
TEXAS STATUS: None

DISTINGUISHING FEATURES. The Jollyville Plateau, Georgetown, and Salado Salamanders share many morphological features, including

Figure 7.17. The Jollyville Salamander (left) and the Georgetown Salamander (right) share many characteristics. Photographs by Piershendrie (left) and Georgetown Salamander (right), Creative Commons.

broad heads, bluntly rounded snouts, and maximum lengths of 2 inches. Their dorsal coloration is grayish brown, their ventral surface is pale translucent, and all retain feathery gills throughout their lives. The Jollyville Plateau (fig. 7.17, left) and Georgetown Salamanders (fig. 7.17, right) each possess 16 costal grooves (vertical indentations) between their limbs, whereas the Salado Salamander may have 15 or 16. Tails in the Jollyville Plateau and Georgetown Salamanders tend to be long, and the poorly developed dorsal and ventral fins are golden yellow at the base, cream colored to translucent toward the outer margin, and mottled with pigment. Unlike the Jollyville Plateau Salamander, the Georgetown Salamander has a distinct dark border along the lateral margins of the tail fin. The terminal portion of the Salado Salamander's tail generally has a well-developed fin on top, but the bottom tail fin is weakly developed (fig. 7.18). Eyes are reduced in the Salado Salamander compared to those of other Central Texas *Eurycea* species. Like the Jollyville Plateau Salamander, recently discovered cave-adapted forms of the Georgetown Salamander have reduced eyes and pale coloration.

SPECIFIC DISTRIBUTION AND HABITAT.
Each species inhabits areas in and around spring openings, spring runs, and pools of water discharged from the northern segment of the Edwards Aquifer. These salamanders also move into caves or other underground areas with underlying groundwater. Some Jollyville Plateau Salamander populations rely on water from geologic formations and alluvial aquifer sources not considered part of the Edwards Aquifer. In addition to moving between surface and subsurface habitats, these salamanders also use interstitial spaces as foraging habitat and protection from predators.

The Jollyville Plateau Salamander occurs in over 100 surface sites and about 16 cave sites in the Jollyville Plateau and Brushy Creek areas of the Edwards Plateau in northern Travis and southern Williamson Counties. The Georgetown Salamander is known from 18 springs along five tributaries to the San Gabriel River and from two caves in Williamson County. Some sites, however, have not been surveyed in recent years. The Salado Salamander occupies four spring sites near the

Figure 7.18. The Salado Salamander. Photograph by Pete Diaz, US Fish and Wildlife Service.

village of Salado in Bell County, and an additional site farther upstream on Salado Creek.

FEDERAL DOCUMENTATION (JOLLYVILLE PLATEAU SALAMANDER):

Listing: 2014. Federal Register 79:10235–10293
Critical Habitat: 2013. Federal Register 78:51327–51379
Recovery Plan: In development

FEDERAL DOCUMENTATION (GEORGETOWN SALAMANDER):

Listing: 2014. Federal Register 79:10235–10293
Critical Habitat: 2012. Federal Register 77:50768–50853 (Proposal)
Recovery Plan: In development

FEDERAL DOCUMENTATION (SALADO SALAMANDER):

Listing: 2014. Federal Register 79:10235–10293
Critical Habitat: 2012. Federal Register 77:50768–50853 (Proposal)
Recovery Plan: In development

Barton Springs Segment of the Edwards Aquifer

Barton Springs Salamander,
Eurycea sosorum
FEDERAL STATUS: Endangered
TEXAS STATUS: Endangered

Austin Blind Salamander,
Eurycea waterlooensis
FEDERAL STATUS: Endangered
TEXAS STATUS: Endangered

DISTINGUISHING FEATURES. Adult Barton Springs Salamanders are approximately 2.5 to 3 inches in length and have reduced eyes and elongate, spindly limbs. The head is broad and deep in lateral view, and the snout appears somewhat truncate when viewed from above. The slight "shovel" appearance of the Barton Springs Salamander's snout and its reduced eyes distinguish it from surface-dwelling Central Texas *Eurycea* species (fig. 7.19). Three bright red, feathery gills occupy each side of neck just behind the head. The mottled color pattern on its upper body varies from light to dark brown, purple, reddish brown, yellowish cream, or orange. The underside of the body is cream colored and often translucent, so that some internal organs and developing eggs in females are visible. There are 15 costal grooves between the limbs. Morphological characteristics that distinguish the Austin Blind Salamander from the Barton Springs Salamander include eyespots covered by skin instead of lenses, an extended snout, fewer costal grooves (12), and pale to dark lavender coloration (fig. 7.20).

Figure 7.19. The Barton Springs Salamander. Photograph by Lisa O'Donnell.

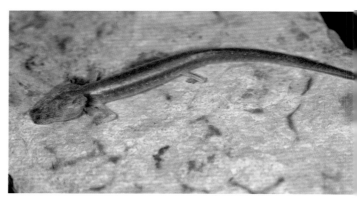

Figure 7.20. The Austin Blind Salamander. Photograph by Dee Ann Chamberlain.

SPECIFIC DISTRIBUTION AND HABITAT. These species occur in the Barton Springs segment of the Edwards Aquifer. The Barton Springs Salamander has been found in the four springs that constitute the Barton Springs system within the city of Austin's Zilker Park. Salamanders believed to be the Barton Springs Salamander have been discovered at several other springs and wells in Travis and Hays Counties, Texas, but genetic analyses of these individuals are pending. The Austin Blind Salamander has been found at only three of the four spring outlets in Zilker Park, where it co-occurs sympatrically (in the same location) with the Barton Springs Salamander. The Austin Blind Salamander is rarely seen and is believed to occupy the underground aquifer more so than the Barton Springs Salamander. In addition to moving between surface and subsurface habitats, both salamanders also use the interstitial spaces as foraging habitat and protection from predators and may also be found in aquatic plants, leaf litter, and woody debris.

FEDERAL DOCUMENTATION (BARTON SPRINGS SALAMANDER):
Listing: 1997. Federal Register 62:23377–23392
Critical Habitat: Not designated
Recovery Plan: US Fish and Wildlife Service. 2005.
Barton Springs Salamander recovery plan (amended 2016 to include the Austin Blind Salamander). Albuquerque, NM. 144 pp.

FEDERAL DOCUMENTATION (AUSTIN BLIND SALAMANDER):
Listing: 2013. Federal Register 62:23377–23392
Critical Habitat: 2013. Federal Register 78:51327–51379
Recovery Plan: US Fish and Wildlife Service. 2005.
Barton Springs Salamander recovery plan (amended 2016 to include the Austin Blind Salamander). Albuquerque, NM. 144 pp.

Southern Segment of the Edwards Aquifer

San Marcos Salamander, *Eurycea nana*
FEDERAL STATUS: Threatened
TEXAS STATUS: Threatened

Texas Blind Salamander, *Eurycea rathbuni*
FEDERAL STATUS: Endangered
TEXAS STATUS: Endangered

DISTINGUISHING FEATURES. Adult San Marcos Salamanders are 1.5 to 2 inches in length and have a small, slender, uniformly reddish-brown body. They alter their color from light tan to dark brown in accordance with the surrounding substrate. Along each side of the dorsum is a row of small, pale spots (fig. 7.21). The underside of the body is cream colored and often translucent, so that some internal organs and developing eggs in females are visible. The head is narrow and converges slightly toward the rounded snout. Compared to those of other Central Texas *Eurycea* species, the eyes are moderate in size and are surrounded partially or completely by a dark ring. The gills are well developed and highly pigmented. The slender tail has a well-developed dorsal fin. The San Marcos Salamander may have 16 or 17 costal grooves between its limbs.

Figure 7.21. The San Marcos Salamander. Photograph by Ryan Hagerty, US Fish and Wildlife Service.

Figure 7.22. The Texas Blind Salamander. Photograph by Ryan Hagerty, US Fish and Wildlife Service.

The Texas Blind Salamander's distinctive cave-adapted morphology includes unpigmented skin that appears smooth and translucent white; a broad, flattened head; thin, elongate limbs; and reduced, nonfunctional eyes visible only as two small dark spots deep within the skin (fig. 7.22). The prominent, well-developed gills are typically bright red and located behind the jaws. There are 12 costal grooves between the limbs. The long tail, which accounts for about half of the total body length, is wide and laterally compressed, with dorsal and ventral fins that taper at the end. In captivity, this species can grow up to 5 inches in length.

SPECIFIC DISTRIBUTION AND HABITAT.
The San Marcos Salamander inhabits spring-fed headwaters of the San Marcos River in Hays County, Texas. It has been found in all spring openings in Spring Lake and in the gravel and rocks below Spring Lake dam. These headwaters receive abundant spring flows through numerous fissures from the underlying Edwards Aquifer. The San Marcos Salamander most commonly occupies cobble and gravel substrates where a plentiful food supply is provided in aquatic moss (e.g., *Amblystegium* spp.) and filamentous algae communities. These aquatic macrophytes occur at depths of 3 to 10 feet within Spring Lake.

The Texas Blind Salamander is known only from springs and aquifers in the San Marcos area of the Edwards Aquifer. Little is known about the Texas Blind Salamander's underground habitat except that it occupies oxygen-rich, water-filled caves with near-constant temperatures.

FEDERAL DOCUMENTATION (SAN MARCOS SALAMANDER):
Listing: 1980. Federal Register 45:47355–47364
Critical Habitat: 1980. Federal Register 45:47355–47364
Recovery Plan: US Fish and Wildlife Service. 1996. San Marcos and Comal and associated aquatic ecosystems (revised) recovery plan. Albuquerque, NM. 134 pp.

FEDERAL DOCUMENTATION (TEXAS BLIND SALAMANDER):
Listing: 1967. Federal Register 32:4001
Critical Habitat: Not designated
Recovery Plan: US Fish and Wildlife Service. 1996. San Marcos and Comal and associated aquatic ecosystems (revised) recovery plan. Albuquerque, NM. 134 pp.

And what is there to life if a man cannot hear the lonely cry of a whippoorwill or the arguments of the frogs around a pond at night? — ATTRIBUTED TO CHIEF SEATTLE (1854)

The huge expanse known as Texas is revered by its inhabitants, partially because of its history, opportunity, and affluence, but also because its landscape is so varied. Across its nearly 800-mile width, biogeographers have identified 10 distinct natural regions by differences in vegetation, which reflect variation in soils, rainfall, and elevation (chapter five). From west to east, the landscape transitions from a dry desert studded with rugged mountains and incised with deep canyons, through grasslands punctuated by stands of junipers and oaks, to forests of tall pines and swampy wetlands. Extensive prairies carpet the Panhandle, North Texas, and the Coastal Plain.

Within each of these terrestrial ecosystems, human intrusions have taken their toll on the populations of many organisms. The Texas Parks and Wildlife Department lists many species as threatened or endangered because habitat alterations have reduced their numbers, condensed their distributional range, or created widely separated subpopulations. Several terrestrial species, such as the Texas Kangaroo Rat and the Texas Tortoise, have such low numbers and restricted distributions that they should qualify for federal protection. Accounts of the terrestrial fauna protected by the Endangered Species Act are provided in this chapter.

American Burying Beetle,
Nicrophorus americanus

J. CURTIS CREIGHTON
AND BRIAN R. CHAPMAN

FEDERAL STATUS: Endangered
TEXAS STATUS: Endangered

Burying beetles are members of a unique beetle guild that locate, defend, and bury vertebrate carcasses, which are used as a food source for their young. For reasons that remain unclear, the American Burying Beetle, *Nicrophorus americanus*, the largest member of the carrion beetle guild, has disappeared from approximately 90 percent of its former range, which once stretched from the Great Plains to the Atlantic Coast. The remaining populations exist on the periphery of the species' historical range. Although a remnant population was known for many years just across the Red River in southeastern Oklahoma, the species was rediscovered in northeastern Texas in 2003. The American Burying Beetle was once widely distributed in Texas, but the current limits of its range in the state are unknown.

DESCRIPTION. Adult American Burying Beetles vary in length from 1.0 to 1.4 inches and are approximately 0.4 inches wide. Each shiny black beetle is adorned with conspicuous orange spots—two irregularly shaped spots embellish each wing cover (elytra) and smaller orange-red

Figure 8.1. An American Burying Beetle is testing a carcass on sandy soil. The chemosensitive tufts on the antennae aid in carcass location. Photograph by Cindy Maynard, US Fish and Wildlife Service.

Figure 8.2. The most recent records of the American Burying Beetle in Texas are from Lamar and Red River Counties. Map by Brian R. Chapman.

of American Burying Beetles have been found in Arkansas, Nebraska, South Dakota, Kansas, and northeastern Texas. In Texas the beetle was found only in Lamar and Red River Counties, but it has not been seen there in several years (fig. 8.2).

HABITAT. Throughout its distributional range, the beetle inhabits wet meadows, grasslands, shrublands, oak-hickory forests, bottomland forests, and partially forested loess canyons. The use of several landscape types may indicate that American Burying Beetles have a broader niche than many other *Nicrophorus* species. Although the species may reside and feed in different habitats, it exhibits preferences for breeding habitat. Successful breeding occurs in grasslands, areas with moderate shrub cover, and forests with substantial leaf litter, but all possess deep, soft soils. Both locations in Texas are within the northernmost extension of the Post Oak Savanna ecoregion where upland forests of Post and Blackjack Oaks are punctuated by grasslands and bottomland hardwood forests in riparian corridors (fig. 8.3). Sandy loams occur on uplands, and clay loams are found in bottomlands.

NATURAL HISTORY. When animals die, aromatic chemicals associated with decomposition are released almost immediately. When temperatures exceed 60°F, adult American Burying Beetles search each night for decaying vertebrates. Chemosensory organs on the antennae detect the chemical signals of decay from distances of up to 2 miles, and the beetles fly to the carcass. Upon landing, the beetles assess the size and weight of the carcass by walking around it and burrowing underneath it. Rodent and bird carcasses weighing between 1.8 and 7.2 ounces are optimal for brood rearing.

Since vertebrate carcasses are an unpredictable resource in both space and time, several beetles often arrive at a carcass and compete for "ownership" of the remains. Eventually, a dominant pair drives off the other competitors. Should a male arrive alone, he climbs to a high point on or near the carcass and releases pheromones to attract a mate.

Once a male-female pair forms, the two beetles sometimes crawl under the remains, turn over onto

blobs adorn the head and mustache-like frons. The key diagnostic feature of the species is the large, orange-red splotch covering most of the pronotum (the rounded plate covering the midsection between the head and wings)—no other North American species of *Nicrophorus* shares this characteristic (fig. 8.1). Males can be distinguished from females by the marking on the frons—the orange spot is rectangular in males and triangular in females. Large antennae originate on the head and terminate in obvious orange clubs.

DISTRIBUTION. Prior to 1920, the distributional range of the American Burying Beetle extended throughout the eastern half of North America, stretching from the prairies and plains east of the Rocky Mountains to the Atlantic Seaboard. When it was placed on the list of endangered and threatened species in 1989, only two populations of the American Burying Beetle were known to exist—one on Block Island off the Rhode Island coast and the other in two counties of southeastern Oklahoma. Since that time, isolated populations

Figure 8.3. American Burying Beetles live in habitats in Texas where oaks and grassy openings intermingle on sandy soils. Photograph by Brian R. Chapman.

their backs, and slowly push the body to a location with suitable soil for burial. The process of carcass burial begins as the pair circles the remains while displacing soil outward and upward with their head and legs. Roots are chewed through and the carcass slowly settles into a depression created by the circling excavators. Before the night is over, the carcass, now below ground level, is covered with a layer of soil, reducing exposure of the carrion to scavengers and preventing flies, which are active during the day, from laying eggs on the remains. After burial, the beetles continue to circumnavigate the carcass, molding it into a ball, removing its hair or feathers, and coating it with oral and anal secretions that inhibit fungal and bacterial growth. Adjacent to the carcass, a brood chamber is hollowed out, and after mating, the female lays eggs in small tunnels surrounding the chamber. The eggs usually require several days to hatch,

depending on temperature. Some offspring are eaten by the parents, ensuring sufficient food for the remaining young.

In a series of behaviors unusual among beetles, parents remain with the carcass and feed the young by regurgitating food directly into their mouths. After developing the ability to feed themselves, the larvae often continue to beg for regurgitated carrion. Both parents guard the brood and work on the carcass to keep it from decaying. The male departs first, about a week after the eggs hatch, but the female lingers at the carcass for up to 14 days, when the larvae disperse into the soil to pupate.

After a metamorphosis requiring approximately 45 days, young beetles emerge as teneral adults (immature beetles incapable of breeding). Another 21-day period is required for the beetles to achieve sexual maturity. Some will breed before the summer is over. Most adult beetles, which live for only a year, spend the winter underground and emerge the following spring to mate once before they die.

REASONS FOR POPULATION DECLINE AND THREATS. Some suggest that the population decline of the American Burying Beetle represents "one of the most disastrous declines of an insect's range ever recorded." Many hypotheses for the decline have been proposed, but no definitive cause has ever been determined. Various researchers have suggested pesticide use, artificial lighting, pathogens, elimination of forests, and regrowth of forests as contributing factors. A combination of three hypotheses provides the most reasonable explanation for the disappearance of this species: (1) reductions of vertebrate populations providing optimally sized carrion; (2) increased numbers of mesopredators and scavengers amplifying competition for carrion; and (3) competition among *Nicrophorus* species for suboptimally sized carrion after optimally sized carrion became less available. Within the beetle's historical range, many bird species that once provided optimally sized carcasses for American Burying Beetle broods experienced population declines or, in the case of the Passenger Pigeon, extinction. While the carrion base was changing, forests were being fragmented into smaller plots.

Fragmented landscapes and the elimination of large predators favored increases in vertebrate scavenger populations (Gray Foxes, Raccoons, Striped Skunks, etc.), which compete with burying beetles for carcass resources. Forced to use slightly smaller carcasses, American Burying Beetles may have entered into direct competition with cogeneric beetles for suboptimal carcasses. Several sympatric congeners could be formidable competitors.

RECOVERY AND CONSERVATION. The recovery plan, published prior to discovery of additional populations, focused on the Block Island population. The five-year review completed in 2008 recommended revision of the plan to establish new recovery goals and objectives in light of the expanded distribution. Until a new recovery plan is available, state and federal governments are responsible for protecting beetle habitats on public lands. The US Fish and Wildlife Service established a permitting process and trapping protocols for surveys to determine where American Burying Beetles occur.

American Burying Beetles are raised in captivity in several locations, and the St. Louis Zoo maintains a large population at its Center for American Burying Beetle Conservation. In 2012, zoo-bred beetles from the center were reintroduced into a preserve in southwestern Missouri owned by The Nature Conservancy. This is a "nonessential experimental population," which means that (1) it is not essential for the survival of the species, and (2) there is flexibility in the rules so that farming and other activities of private landowners are not affected. Experimental populations also exist on Penikese Island and Nantucket Island, Massachusetts.

FEDERAL DOCUMENTATION:
Listing: 1989. Federal Register 54:29652–29655
Critical Habitat: Not designated
Recovery Plan: US Fish and Wildlife Service.
    1991. American Burying Beetle (*Nicrophorus americanus*) recovery plan. Newton Corner, MA. 80 pp.

Houston Toad, *Anaxyrus houstonensis*

PAIGE A. NAJVAR

FEDERAL STATUS: Endangered
TEXAS STATUS: Endangered

The Houston Toad was one of the first amphibians federally listed as an endangered species.

DESCRIPTION. Houston Toads are generally brown and speckled, although individual coloration may vary from light brown or red to gray with dark brown or black spots (fig. 8.4). The underside is usually pale with small dark spots. The legs are banded with dark pigment, and two dark bands extend from each eye down to the mouth. A white stripe that varies in pigmentation density extends down the middle of the back, but it may be absent in some individuals. Males and females are similar in appearance, although females are typically larger and bulkier than males. Males have dark throats, which appear bluish when distended. Adult Houston Toads are 2 to 3 inches long and, like all toads, are covered with raised patches of skin that resemble warts.

DISTRIBUTION. The Houston Toad is endemic to 13 counties in east-central Texas, including Austin, Bastrop, Brazos, Burleson, Colorado, Fort Bend, Harris, Lavaca, Lee, Leon, Liberty, Milam, and Robertson (fig. 8.5). The species is likely extirpated from Brazos, Fort Bend, Harris, and Liberty Counties.

Figure 8.4. An adult Houston Toad. Photograph by Paige A. Najvar.

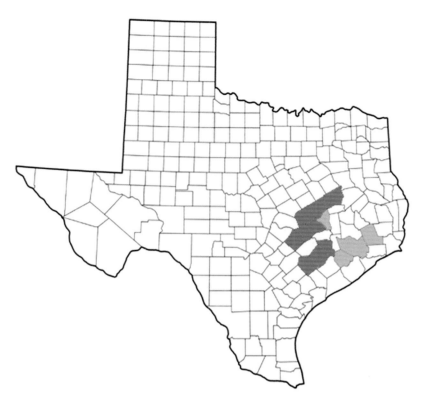

Figure 8.5. The Houston Toad can still be found in nine Texas counties (red), but it no longer occurs in four other counties (yellow) that were part of its recent range. Map by Brian R. Chapman.

Figure 8.6. Shallow breeding ponds used by the Houston Toad often dry during the summer months. The toads also inhabit pine-oak woodlands, as seen in the background. Photograph by Paige A. Najvar.

HABITAT. The Houston Toad inhabits rolling uplands with mixed-canopy pine and oak forests with sandy soils and native herbaceous ground cover. In these habitats the toads breed in small pools of water and ephemeral ponds (fig. 8.6). For successful breeding and emergence of juveniles, water must persist for at least 60 days. Adjacent uplands support adults year-round and provide patch connectivity outward from the ponds for juvenile dispersal.

NATURAL HISTORY. The Houston Toad is an "aggregative" and "explosive" breeder that historically appeared in large numbers at breeding ponds where the males chorused to attract females and competed with each other to breed over a few nights. Female Houston Toads reach sexual maturity in two years, as opposed to one year in males. Chorusing can begin as early as January, and breeding typically takes place from late January to June, with a peak in reproductive activity in March, depending on weather conditions. Breeding is believed to be triggered in part by rainfall and warm nighttime temperatures. Males clasp females with their front legs, a part of the mating process known as amplexus, and fertilize eggs externally as they are deposited by the female. In a single breeding effort, one female can deposit several hundred eggs, which are laid in strings (egg strands) in the pond. The eggs hatch into tadpoles that metamorphose into juveniles approximately 60 to 65 days after deposition. Following metamorphosis, juveniles move into terrestrial habitats where they grow and develop into adults.

The Houston Toad appears to have evolved in a metapopulation (a dynamic system of populations interconnected by emigration and immigration). The key to the Houston Toad's metapopulation connectivity is juvenile dispersal, which provides linkages among subpopulations. After they emerge from their natal ponds, most juvenile Houston Toads disperse across terrestrial habitat to colonize other breeding sites. Adult Houston Toads typically remain within 656 feet of their breeding pond during the postbreeding season.

Houston Toads feed on a variety of insects and other invertebrates that reside within the herbaceous ground cover of the forest floor. Tadpoles sometimes ingest pine pollen and the jelly envelopes of recently hatched Houston Toad eggs. Tadpoles remain on the bottom of ponds during the day and feed on material attached to the vegetation in and along the pond's edge at night.

REASONS FOR POPULATION DECLINE AND THREATS. Extensive loss, fragmentation, and disturbance of forested areas have reduced the Houston Toad's habitat throughout its range. Sources of habitat loss or disturbance include expanding urbanization, conversion of woodlands to agricultural use, logging, mineral production, alteration of watershed drainages, wetland degradation or destruction, fire suppression, and catastrophic wildfire.

Livestock wading and Feral Pig use can destroy habitat around the perimeter of a Houston Toad breeding pond and result in degraded water quality (from nitrogenous wastes such as urine and manure) and an overall adverse environment for amphibian egg and tadpole development. Houston Toad eggs and juveniles are also crushed by livestock trampling and Feral Pig wallowing at breeding ponds. Red Imported Fire Ants prey on newly metamorphosed Houston Toads and eliminate the invertebrate species that make up the Houston Toad's food supply.

Drought conditions, an additional stressor for the Houston Toad, reduce soil moisture and decrease pond levels, resulting in desiccation of Houston Toad adults, juveniles, and eggs. As drought severity increases, predators, livestock, and Feral Pigs tend to concentrate at the remaining breeding sites, further exacerbating habitat destruction and Houston Toad mortality. Given their already low populations and other ongoing threats, Houston Toad populations have difficulty rebounding after drought conditions subside.

Although breeding aggregations of several hundred individuals were observed at ponds as recently as the 1980s, choruses of fewer than 10 individuals have been more typical throughout the Houston Toad's range since 2000. These small populations isolated by habitat loss and fragmentation escalate the risk of extinction by limiting the species' ability to naturally recolonize an area following a local extirpation event. Small populations may also suffer a loss of genetic diversity over time, reducing their ability to evolve in response to changing environmental conditions.

RECOVERY AND CONSERVATION. The Houston Toad evolved in large, well-connected populations spanning its range. Habitat fragmentation and loss have resulted in considerably smaller populations that are more geographically isolated. Therefore, habitat management for Houston Toad conservation is directed toward improving juvenile and adult survivorship as well as increasing dispersal and connectivity among local subpopulations.

Houston Toad extinction probabilities greatly decrease with increased juvenile survivorship, low adult mortality, and improved dispersal between habitat patches. Thus, captive propagation and population supplementation efforts aimed at increasing the number of Houston Toad juveniles and preventing the immediate extinction of this species in the wild have been in place since 2007, primarily within areas designated as critical habitat within Bastrop County, Texas. There are plans to expand this program across the Houston Toad's range.

Landowner cooperation is critical to implementing habitat management and restoration efforts throughout the Houston Toad's range. The US Fish and Wildlife Service is partnering with various state and federal agencies, local governments, and nongovernmental organizations to engage private landowners in Houston Toad conservation through various incentive programs.

FEDERAL DOCUMENTATION:

Listing: 1970. Federal Register 35:16047–16048

Critical Habitat: 1970. Federal Register 43:4022–4026

Recovery Plan: US Fish and Wildlife Service. 1984. Recovery plan for the Houston Toad (*Bufo houstonensis*). Albuquerque, NM. 75 pp.

Louisiana Pinesnake,
Pituophis ruthveni

WILLIAM I. LUTTERSCHMIDT, D. CRAIG RUDOLPH, AND JOSH B. PIERCE

FEDERAL STATUS: Candidate
TEXAS STATUS: Threatened

The Louisiana Pinesnake is the rarest Texas snake species. Its secretive habits make it difficult to find unless it is encountered on open paths and roads. The snake has a limited distribution, occurring almost exclusively in open, frequently burned pine forests with habitat characteristics similar to those where the endangered Red-cockaded Woodpecker

occurs. Louisiana Pinesnakes are terrestrial (not arboreal) and spend most of their time in or near sandy burrows made by pocket gophers.

DESCRIPTION. The Louisiana Pinesnake resembles the Bullsnake but is distinguished from the latter by having 28 to 42 dark body saddles and 6 to 13 tail spots. At midbody, there are usually 31 (range 27 to 33) rows of scales. The ground color is white or tan (rarely yellow), the midbody is usually pale, and the nape is mottled (fig. 8.7). Other distinguishing characteristics include a smaller head, four prefrontal scales on the forecrown, and a high rostral scale (at the tip of the snout). A postorbital bar is prominent on some specimens. Adults are large bodied, ranging from 48 to 60 inches in length, but a maximum size of 70.5 inches can be attained.

DISTRIBUTION. The Louisiana Pinesnake once occurred in 13 East Texas counties and nine Louisiana parishes. The most recent records, now more than a decade old, are restricted to five counties, including the southern portion of Sabine National Forest and adjacent private lands (Sabine and Newton Counties) and the southern portion of the Angelina National Forest (Angelina, Jasper, and Tyler Counties) (fig. 8.8). The species is probably now extirpated in Texas.

NATURAL HISTORY. Although much information is available concerning the natural history of some

Figure 8.7. This Louisiana Pinesnake in its characteristic S-shaped defensive posture is hissing to discourage further harassment. Photograph by Brendan Kavanagh.

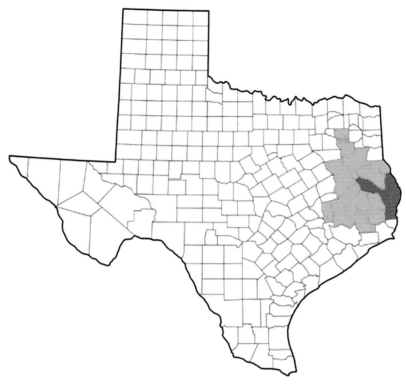

Figure 8.8. The Louisiana Pinesnake may now occupy only three counties (red) within its former range (yellow). Map by Brian R. Chapman.

Pituophis species, much less is known about the Louisiana Pinesnake. The behavior and ecology of a morphologically similar pinesnake occurring in the New Jersey Pine Barrens provide some basic information. Although the behavior of individuals may vary, *Pituophis* species are famous for their defensive posture—when encountered, they raise their head above their body in an S-shaped coil, vibrate their tail, and hiss loudly. The intimidating sound is created when the snake inflates itself to capacity and exhales, forcing the air over the glottis near the front of its mouth. This behavior and the saddle-shaped markings likely contribute to increased mortality because pinesnakes often are mistaken for rattlesnakes. Some accounts describe *Pituophis* as a nervous creature, potentially striking and even biting when first handled. When threatened, pinesnakes quickly retreat to gopher burrows.

The prey of Louisiana Pinesnakes consists mostly of small mammals, including Baird's Pocket Gophers (over 50 percent of the diet), which are common in the sandy soils where Louisiana Pine-snakes occur. Turtle eggs have also been recorded, but captive animals usually refused insects and lizards when offered.

Little is known about reproduction or the reproductive success of the Louisiana Pinesnake in the wild. Captive females lay one to five eggs that measure about 4.6 inches in length and 1.3 inches in width. The large sizes of eggs and hatchlings are remarkable compared to those of other members of the genus. The large eggs may facilitate the rapid attainment of a size capable of feeding on pocket gophers. Incubation is between 58 and 66 days, and hatchlings are between 17.3 and 22.0 inches long. Nest sites for the Louisiana Pinesnake remain unknown.

HABITAT. The Louisiana Pinesnake seems to prefer open pine woods where Longleaf Pine is the dominant vegetation. Longleaf Pine forests are associated with sandy, well-drained soils that support savanna-type habitats with a moderate to sparse midstory under a canopy of pines and a well-developed herbaceous understory dominated by grasses (fig. 8.9). Adjacent Blackjack Oak woodlands and Shortleaf Pine and Post Oak forests may also serve as suitable habitat. Within such habitats, the snakes frequent the burrow systems of Baird's Pocket Gophers to forage, find shelter, and escape from fire.

Figure 8.9. The Louisiana Pinesnake inhabits sandy soils in the forested habitats of East Texas. Photograph by Brian R. Chapman.

REASONS FOR POPULATION DECLINE AND THREATS. Development of agricultural lands, urban expansion, road construction, and increased vehicle traffic play major roles in the continuing decline of the Louisiana Pinesnake. However, the initial impact on this species began with extensive harvests of Longleaf Pine, which were largely complete by the 1920s. The original Longleaf Pine forests were replaced with plantations of faster-growing Loblolly and Slash Pines. These managed forests—dense stands with sparse and poorly structured understory plant communities—altered the landscape and the snake's preferred habitat. Longleaf Pine forests cover less than 3 percent of their original range in Louisiana and Texas. Suppression of natural fire represents a significant threat, decreasing both the quantity and quality of habitat available. Longleaf Pine savannas are fire-climax communities, adapted to the occurrence of frequent low-intensity ground fires. In the absence of periodic fires, upland pine savanna ecosystems rapidly develop a dense midstory that suppresses herbaceous understory growth. Louisiana Pinesnakes are well adapted to fire—they commonly move into pocket gopher burrows and survive low-intensity fires. Habitat fragmentation, lack of demographic viability, and low genetic diversity are increasingly of concern. Human predation and collection for the pet trade also constitute potential threats.

RECOVERY AND CONSERVATION. A candidate conservation agreement, approved in 2004, outlined conservation and management measures designed to protect the Louisiana Pinesnake within federal properties in Louisiana and Texas. The USDA Forest Service, US Army, US Fish and Wildlife Service, and Texas Parks and Wildlife Department agreed to protect the species within the boundaries of the Angelina and Sabine National Forests of Texas and the Kisatchie National Forest and Fort Polk Military Reservation in Louisiana. Habitat restoration efforts will be similar to those required for managing habitats for Red-cockaded Woodpeckers—namely, regular prescribed burning, thinning, and replanting of Longleaf Pine forests. Beginning in 2010, captive-bred animals from zoos were released into large tracts of restored habitat on the Catahoula District of the Kisatchie National Forest. A similar effort is planned for Texas.

FEDERAL DOCUMENTATION:
Listing: 2014. Federal Register 79:72449–72497
Critical Habitat: Not designated
Recovery Plan: Not developed

Mexican Long-nosed Bat,
Leptonycteris nivalis

LOREN K. AMMERMAN

FEDERAL STATUS: Endangered
TEXAS STATUS: Endangered

Of the 33 species of bats found in Texas, only 2 forage on nectar and pollen rather than on insects. One of these, the Mexican Long-nosed Bat, *Leptonycteris nivalis*, roosts in caves and engages in long-distance movements coinciding with the blooming of agaves and cacti in the deserts of northern Mexico, southern New Mexico, and western Texas. Because the species migrates seasonally and occupies isolated habitats that are challenging to census, the current population size is unknown. However, the loss of food sources, disturbances at roosts, extermination by humans, and conversions of desert habitats to agricultural lands has caused significant population declines. The Mexican Long-nosed Bat is also listed as endangered under Mexico's endangered species law.

DESCRIPTION. The average total length of the bat is 3.25 inches. A tiny, inconspicuous tail containing three vertebrae is located at the base of the uropatagium (the membrane between the hind legs). Individuals vary slightly in color from dark gray to yellowish gray, but the underside appears lighter in color because the hairs are white tipped. The ears are short, the muzzle is long and somewhat pointed, and the nose bears a prominent, triangular nose leaf (fig. 8.10). The bat has a long tongue with hairlike papillae at the tip. The Mexican Long-tongued Bat, which is similar

Figure 8.10. A heavy dusting of pollen garnishes the head of this Mexican Long-nosed Bat. Photograph by Loren K. Ammerman.

in appearance, differs in having a longer, narrower muzzle and a distinct tail.

TAXONOMIC COMMENTS. The Mexican Long-nosed Bat was originally described in 1860 from a specimen collected high on Mount Orizaba in Mexico. The name *nivalis* (snow) indicates that the specimen was captured at the snow line. No subspecies are recognized. Because the Lesser Long-nosed Bat was once considered a subspecies of the Mexican Long-nosed Bat, information about the Lesser Long-nosed Bat has sometimes been mistakenly ascribed to the Mexican Long-nosed Bat.

DISTRIBUTION. Although it occurs throughout most of central Mexico and as far south as the states of Guerrero and Puebla, the Mexican Long-nosed Bat is known in the United States from only a few locations in Texas and from the Animas and Big Hatchet Mountains in New Mexico. Females of this species roost seasonally in Mount Emory Cave in the Chisos Mountains of Big Bend National Park (BBNP) in Brewster County, Texas, and a male specimen has been captured in the Chinati Mountains in Presidio County, Texas (fig. 8.11).

The population of Mexican Long-nosed Bats inhabiting Mount Emory Cave each summer fluctuates annually but has averaged 2,400 bats (1,790–3,238 bats) over the last eight years. The fluctuations could be related to the abundance of food resources in Mexico or Texas.

Seasonal movements to ensure an uninterrupted food supply are related to the blooming periods of the agaves and cacti on which the bat feeds. Migration northward from central Mexico follows the progression of spring flowering. The bats reach BBNP in mid-May, about a month after the agaves begin flower production. When agaves stop blooming in late summer or early fall, the bats move southward to overwinter in the Central Valley of Mexico, where mating occurs and many species of flowering plants provide sustenance.

HABITAT. In Texas, although Mexican Long-nosed Bats are occasionally encountered in desert and semidesert habitats at lower elevations, these bats apparently prefer mountainous pine-oak habitats at elevations of 4,900–7,500 feet. Caves provide diurnal roosts. High in the Chisos Mountains, Mount Emory Cave, the only known day roost in Texas, is a cool cave with a constant breeze. There, the bats cluster in tight groups of sometimes several thousand individuals positioned on high ceilings in passageways and inaccessible areas deep within the cave. Clusters can be easily detected by the yellowish, semiliquid droppings that accumulate on the cave floor beneath a roost; a distinctive sweet-smelling aroma emanates from the musty deposits.

NATURAL HISTORY. Many aspects of Mexican Long-nosed Bat life history are known only from brief observations. Detailed ecological studies are underway.

The Mexican Long-nosed Bat emerges from day roosts less than an hour after sunset to feed on nectar and pollen. Occasionally shuttling up to 24 miles between roosting and feeding locations, the bats often forage in small groups, a behavior that likely facilitates navigation, reduces predation risk, and increases the likelihood of discovering plants at the peak of nectar production. The strong and agile fliers are capable of high flight speeds and make rapid vertical ascents and turns to avoid obstacles or predators.

The bat's long muzzle and tongue provide access

Figure 8.11. The only Texas records for the Mexican Long-nosed Bat are from Brewster and Presidio Counties. Map by Brian R. Chapman.

to nectar and protein-rich pollen. It feeds on flowers of at least 21 species of plants, often while hovering. Insects are also occasionally ingested. In the southern part of the bat's range, the Morning Glory Tree is an important food source. In Texas, the Mexican Long-nosed Bat feeds primarily on Havard's Century Plant (fig. 8.12). Because pollen adhering to the fur is carried from one plant to another, the long-nosed bats are regarded as the most significant pollinators of many species of agaves, particularly the species used to produce tequila in Mexico. Although most agaves that are used to produce tequila and other alcoholic drinks are cultivated and vegetatively propagated, wild agave populations depend on long-nosed bats to maintain genetic diversity by cross-pollination. Although agave plants can reproduce asexually by sending out ground shoots from the main stem, the association between the nectar-feeding bats and agaves may be important to the survival of both organisms.

Mating season is from September to February. A cave in Morelos, Mexico, is the only known mating location, and most of the young are born

Figure 8.12. The upturned flowers of Havard's Century Plant enable the Mexican Long-nosed Bat to feed while hovering. Photograph by Brian R. Chapman.

from May to June in Mexico. However, the presence of pregnant and lactating females in Mount Emory Cave indicates that some births likely occur in Texas. Females usually give birth to a single pup.

REASONS FOR POPULATION DECLINE AND THREATS. The population decline is likely related to the loss of roosting sites and decline of food resources. Animals that roost in large aggregations in only a few locations are vulnerable to disturbance, and the Mexican Long-nosed Bat is especially sensitive to human disturbance at the roost. Although Mount Emory Cave is somewhat protected because few people venture up the nearby trail, it is located within a popular area of Big Bend National Park that receives many visitors. No special efforts to prevent disturbance at the roost have been made. Human disturbance at Mexican roosting sites may be more common, especially where uninformed citizens sometimes kill all bats in a cave because they fear vampire bats.

A greater contributor to the population decline might be the elimination of food plants in migratory pathways. Agaves, the bats' most important food resource during their migrations, are regularly harvested, legally and illegally, for the production of tequila, mescal, and pulque. Up to 1.2 million wild agaves are harvested illegally each year by "moonshiners," who produce bootleg mescal and other unlicensed alcoholic beverages. In central Mexico, large areas of natural habitats that once supported agaves and other flowering plants have been converted to croplands or rangelands. Almost no natural vegetation remains after such modifications. When natural habitats are altered, competition for food between the species of long-nosed bats might cause further population declines for one or both species where their distributional ranges overlap.

RECOVERY AND CONSERVATION. Because the species roosts in only a few known locations in the United States and is easily disturbed at roosting sites, the US Fish and Wildlife Service chose not to publish critical habitat rules for the Mexican Long-nosed Bat. To minimize disruption by vandals or curiosity seekers, no descriptions or localities of roosting sites were published, as is normally required for critical habitat descriptions.

The recovery plan for the Mexican Long-nosed Bat does not prescribe criteria for population size and stability indicators but instead provides "interim recovery" objectives. The plan calls for the protection of known roosting sites and foraging habitats within a 24-mile radius of known roosting sites. Scientific work focused on the life history, ecology, and seasonal movements of the species is suggested. A "grassroots" education program targeting communities near known roosting sites to alter misconceptions about bats is also proposed.

According to the recovery plan, the Mexican Long-nosed Bat may be downlisted from endangered to threatened when (1) at least six populations and the essential seasonal roosting and foraging habitats for each are protected, (2) the six populations are maintained for at least 10 consecutive years, and (3) the data indicate that the six populations and their habitats will continue to be maintained. The six populations mentioned in the plan are considered a "tentative number."

Significant contributions to ecological knowledge about the Mexican Long-nosed Bat have been made in both Mexico and the United States since the recovery plan was published. For example, researchers found that large areas of wild agaves must be protected to maintain a stable population at a roost in Nuevo León, Mexico. Thermal imaging technology is enabling investigators to determine the total population of Mexican Long-nosed Bats in Texas, and tiny radio transmitters attached to bats have provided more detailed information about movements and habitat use. More recently, microchips implanted in some bats are being detected as the bats enter and exit their roost cave. This technology is revealing seasonal aspects of cave use and refining our understanding of migratory movements.

Current efforts to protect roosting sites in the United States and Mexico are minimal. Although the population of Mexican Long-nosed Bats appears to have stabilized, the threat of ongoing habitat alteration in Mexico and the vulnerability of roosting sites to human disturbance throughout its range render recovery and downlisting extremely problematic.

FEDERAL DOCUMENTATION:
Listing: 1998. Federal Register 53:38456–38460
Critical Habitat: Not designated
Recovery Plan: US Fish and Wildlife Service. 1994.
    Mexican Long-nosed Bat (*Leptonycteris nivalis*)
    recovery plan. Albuquerque, NM. 98 pp.

Louisiana Black Bear,
Ursus americanus luteolus

CHRISTOPHER COMER

FEDERAL STATUS: Delisted in 2016
STATE STATUS: Threatened

The Louisiana Black Bear is one of 16 recognized subspecies of the American Black Bear and one of 3 subspecies that historically occurred in Texas. The Louisiana Black Bear once ranged widely in southeastern Texas but was eliminated from everywhere but the Piney Woods before 1900, and bears were hunted out of the Piney Woods in the early twentieth century. The US Fish and Wildlife Service listed the Louisiana Black Bear as a threatened species in 1992 when the two populations remaining in Louisiana consisted of as few as 90 total bears. Numbers of bears in Louisiana have since increased, and individuals occasionally cross the Sabine River into East Texas.

DESCRIPTION. The Louisiana Black Bear is a bulky mammal with dense, mostly black fur, a brownish muzzle, and a short tail (fig. 8.13). A few individuals exhibit a white blaze on the chest or cinnamon-tinged pelage. The bears typically measure 2–3 feet tall at the shoulder and 5–6.5 feet in length. Males, which are considerably larger than females, may weigh between 175 and 300 pounds, whereas adult females typically weigh between 90 and 150 pounds. All Black Bears have short, rounded ears and smallish eyes. A longer, narrower skull and larger molar teeth distinguish the relatively small Louisiana subspecies from American Black Bears in other regions.

DISTRIBUTION. With a historical range spanning the continent from Alaska and Canada to northern

Figure 8.13. Green vegetation is a major component of the Louisiana Black Bear diet in the spring. Photograph by Pam McIlhenny, US Fish and Wildlife Service.

Mexico, the American Black Bear is the most widespread and abundant bear species in North America. The historical range of the Louisiana Black Bear included all of Louisiana and adjacent portions of southwestern Mississippi, eastern Texas, and southern Arkansas. The current distribution includes four areas in Louisiana within or adjacent to the Lower Mississippi Alluvial Valley and western Mississippi along the Louisiana border. Although a breeding population of Louisiana Black Bears no longer exists in East Texas, between 2000 and 2016 the Texas Parks and Wildlife Department recorded 24 documented sightings in the region (fig. 8.14).

HABITAT. Like other subspecies of the American Black Bear, the Louisiana Black Bear is an adaptable habitat generalist. In Texas, Louisiana, and Mississippi the species often closely inhabits large contiguous blocks of bottomland hardwood forest. These forests along major rivers and streams are dominated by a mature canopy of mast-producing trees with dense understory thickets of River Cane and similar vegetation.

NATURAL HISTORY. Louisiana Black Bears are best described as opportunistic omnivores. The majority of their diet is plant based, and they depend on hard and soft mast during much of the year. The bears eat green vegetation in the spring and devour soft mast, including Blackberries,

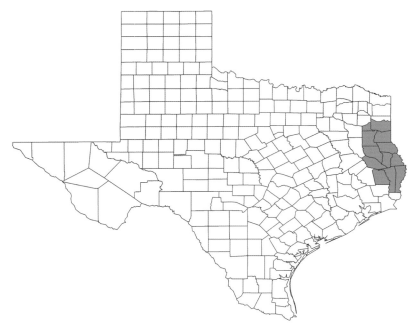

Figure 8.14. Although there are no recent breeding records from the state, Louisiana Black Bears occasionally wander into several East Texas counties from a population in western Louisiana. Map by Brian R. Chapman.

Dewberries, and Wild Grapes, in the summer (see fig. 8.13). Fall consumption of hard mast, primarily oak acorns and hickory nuts, is critical for the accumulation of winter fat reserves. Carrion, insects, and small animals supplement the diet throughout the year.

Louisiana Black Bears do not hibernate but spend a brief period of winter dormancy in dens, even at the extreme southern extent of their range. Cavities in large-diameter trees, particularly Baldcypress and Water Tupelo, are their preferred den sites, but they also repose in dense thickets, brush piles, logging slash, and ground nests.

Black Bears typically reach sexual maturity between three and five years of age. Breeding occurs in the summer, but implantation of the embryo is delayed for several months. When the cubs are born in the den by early February, they are altricial (hairless, helpless, and with eyes closed). Most bears produce two young, but litter size ranges from one to four. Cubs remain with their mother throughout the summer and fall and enter the den with her the following winter. The young disperse after they emerge from the den in the spring. Females produce offspring every other year.

REASONS FOR POPULATION DECLINE AND THREATS. Louisiana Black Bear populations declined as a result of historical unrestricted harvest and habitat loss and fragmentation. Unregulated hunting in Texas continued into the mid-1900s even after bear populations had exhibited widespread decline. By the early twentieth century, extensive logging and conversion to other uses had eliminated about 80 percent of the bottomland hardwood forests throughout the bear's historical range in Texas. Some bottomland forests were replanted with pines, but the remaining forested tracts existed as unconnected fragments.

RECOVERY AND CONSERVATION. Coordinated efforts involving the Louisiana Department of Wildlife and Fisheries, the US Fish and Wildlife Service, the Black Bear Conservation Coalition, and researchers from various universities led to successful recovery of the Louisiana Black Bear in Louisiana. The subspecies was removed from the federal list of threatened species on 11 March 2016. Future conservation efforts in Texas will likely focus on protecting suitable habitats and managing the natural immigration and recolonization by bears from existing populations in Louisiana.

FEDERAL DELISTING DOCUMENTATION: 2016. Federal Register 81:13124–13171

Mexican Gray Wolf, *Canis lupus baileyi*

BRIAN R. CHAPMAN

FEDERAL STATUS: Endangered
TEXAS STATUS: Endangered (Extirpated)

The Gray Wolf no longer occurs in Texas, but it formerly roamed the western two-thirds of the state. Two subspecies inhabited Texas—the Texas Gray Wolf, which is extinct, and the Mexican Gray Wolf (or Mexican Wolf). Two Mexican Gray Wolves killed in the Trans-Pecos by ranchers in 1970 represent the last documented reports of wolves in Texas. Although wolves were extirpated from Texas,

Figure 8.15. The reintroduced Mexican Gray Wolf population in New Mexico could eventually expand its range to reoccupy habitats in West Texas, but the prospects are bleak. Photograph by Jim Clark, US Fish and Wildlife Service.

descendants of captive-bred Mexican Gray Wolves were reintroduced beginning in 1988 into eastern Arizona and southwestern New Mexico, where they continue to roam freely.

DESCRIPTION. The Mexican Gray Wolf resembles a German Shepherd in shape and size. Its pelt is yellowish gray washed with black dorsally, and its tail is short with a blackish tip. The head of some appears shaded with cinnamon, but others are tinted with yellow (fig. 8.15). All gray wolves have long legs and large feet, but the Mexican Gray Wolf is slightly smaller in all dimensions than other subspecies of the Gray Wolf.

DISTRIBUTION. The historical range of the Gray Wolf once covered all of North America except for the southeastern United States, and Central America as far south as Panama. In earlier times, the Mexican Gray Wolf ranged throughout the sparsely populated Trans-Pecos region, and the Texas Gray Wolf occupied the High Plains and Rolling Plains. Soon after ranchers and settlers arrived in North Texas, the Texas Gray Wolf was eradicated—the subspecies was rendered extinct. The Mexican Gray Wolf persisted much longer in the rugged mountains of West Texas and northern Mexico, but it too was eventually eliminated. The Mexican Gray Wolf has been reintroduced to parts of its former range in Arizona and New Mexico and could eventually move back into Texas.

HABITAT. Throughout most of their range, Mexican Gray Wolves likely preferred open grasslands broken by canyons and arroyos. Montane forests and grasslands in West Texas served as a last refuge.

NATURAL HISTORY. The Mexican Gray Wolf travels and hunts in packs of 5–10 members. Each pack consists of a dominant male and a dominant female (an alpha pair), their offspring, and a few nonrelated individuals. After mating, which occurred as early as January in southwestern Texas, the female gives birth to a litter of six pups in a secluded site. As with most canids, the pups are blind and helpless at birth. At the age of about 8–10 weeks, the young are moved to a more open "rendezvous site" where they romp and play over an area of 2 acres or more. While the pups are developing their stature and motor abilities, members of the pack provide food for the mother and her offspring. The young reach adult size by October and begin accompanying the pack during hunting forays.

Their historical diets are not well documented, but those of reintroduced populations suggest that they likely preyed on White-tailed and Mule Deer, Desert Bighorn Sheep, Pronghorn, and occasionally Elk. Smaller mammals such as American Beaver, Collared Peccary, Black-tailed Jackrabbits, rabbits, and rodents were also consumed.

Members of a pack communicate using displays involving facial expressions, howling, other vocalizations (growling and barking), and scent marking. Distinctive howls of each individual assemble the pack and advertise territorial boundaries. The territory size of the Mexican Gray Wolf was never documented, but a Gray Wolf occupies a territory ranging from about 50 to 5,000 acres. Packs of the Mexican subspecies likely traveled widely and maintained large territories in West Texas.

REASONS FOR POPULATION DECLINE. The Mexican Gray Wolf was eradicated because it was responsible for widespread depredation of livestock (sheep, goats, and cattle). Wolves were shot, trapped, and poisoned throughout their range.

RECOVERY AND CONSERVATION. All subspecies of the Gray Wolf in the United States except for a population in Minnesota were listed as endangered in 1976; the Minnesota population was classified as threatened. Since that time, many western populations have been delisted because of recovery. The reintroduced Gray Wolf population in Yellowstone, which expanded rapidly into surrounding areas in Wyoming, Montana, and Idaho, is listed as an experimental population.

The Mexican Gray Wolf Recovery Team prescribed a captive-breeding program that would lead to reintroduction of at least 100 wolves to their historical range. No delisting criteria were proposed because the recovery team had little hope that the subspecies could survive.

Four males and a pregnant female captured in Mexico were the founders of the captive-breeding program. Offspring from the program were released into the Apache National Forest in Arizona, and the population subsequently expanded into New Mexico. By 2014, at least 109 wolves occupied southwestern New Mexico and southeastern Arizona. Because most Gray Wolf populations were being removed from endangered status, the Mexican Gray Wolf was listed separately as endangered in 2015. A federal judge recently ordered the US Fish and Wildlife Service to develop a new recovery plan for the Mexican Gray Wolf by November 2017.

There are no current plans to reintroduce the subspecies into Texas. It is unlikely that wandering individuals will enter the state in the near future.

FEDERAL DOCUMENTATION:

Listing: 2015. Federal Register 80:2488–2512
Critical Habitat: Not designated
Recovery Plan: US Fish and Wildlife Service. 1982.
   Mexican Wolf recovery plan. Albuquerque, NM.
   103 pp.

Red Wolf, *Canis rufus*

BRIAN R. CHAPMAN

FEDERAL STATUS: Endangered
TEXAS STATUS: Endangered (Extirpated)

The distinctive howls of the Red Wolf were no longer heard in Texas after the last individuals were removed in 1973 to initiate a captive-breeding program. Prior to their forced extirpation, a remnant population occupied a few counties and parishes in the Gulf Coast region of southeastern Texas and southwestern Louisiana.

DESCRIPTION. Intermediate in size between a Coyote and a Gray Wolf, a Red Wolf resembles a German Shepherd in size and posture. The coat has some black along the back, but brownish or buff colors dominate the sides and rump, and it has a bushy, black-tipped tail (fig. 8.16, left). The fur on its muzzle, the backs of its ears and legs, and its haunches has a rusty appearance, which lends credence to its common name (fig. 8.16, right).

Some question the validity of the Red Wolf as a distinct species. Recent genetic evidence suggests that the Red Wolf is not a hybrid between the Gray Wolf and the Coyote, but a distinct species that diverged from the Coyote 150,000–300,000 years ago. The species existed historically in southeastern North America where Gray Wolves and Coyotes did not occur.

DISTRIBUTION. The distributional range of the Red Wolf formerly extended from Central Texas eastward through the southern tier of states to Florida and as far north as Pennsylvania. Human intervention (primarily habitat alteration and unregulated hunting) eventually contracted the range of the species to the Gulf Coast region of Texas and Louisiana. Today, Red Wolves exist in the wild only on the Albemarle-Pamlico Peninsula of eastern North Carolina, where they were introduced beginning in the 1980s. The survival of that population is threatened by interbreeding with Coyotes.

Figure 8.16. Although they no longer inhabit the Texas Gulf Coast, descendants of Red Wolves taken from Texas thrive in a similar habitat in North Carolina. Photographs by Steve Hillibrand, US Fish and Wildlife Service.

HABITAT. Prior to their elimination, Red Wolves occupied a variety of habitats, as suggested by their widespread distribution. The journals of William Bartram and John James Audubon indicate that the wolves remained hidden in hardwood forests with dense understories during daylight hours. The last free-ranging Red Wolves in Texas and Louisiana were confined to swamps, dense thickets, and coastal prairies along the Gulf Coast.

NATURAL HISTORY. Red Wolves mate for life, and a dominant pair forms the nucleus of a pack that hunts cooperatively. A pack usually consists of five to eight individuals, most of which are offspring of the dominant pair. At the age of one to three years, young wolves typically leave the pack to find a mate and establish their own pack and territory.

Three to five pups are born in the spring and are raised in dens in hollow trees, culverts, or stream banks. Pack members provide food and protection for the pups. Red Wolves prey on small mammals such as rodents and rabbits but occasionally take larger mammals. Fears that the wolves introduced into North Carolina might harm livestock have proven groundless, but untended pets sometimes disappear.

REASONS FOR POPULATION DECLINE AND THREATS. Early settlers shot, trapped, and poisoned Red Wolves for fear they would prey on livestock. During the same period, large tracts of hardwood forests were cleared, eliminating the species' essential habitats. The resulting habitat modifications favored an invasion of Coyotes, which began to interbreed with the remaining Red Wolves, forming a population composed largely of Coyote × Red Wolf hybrids. Through a process known as introgression (genetic swamping), the hybrids became more similar to Coyotes than to Red Wolves in appearance and behavior. Because the Red Wolf was a vanishing species, it was included on the initial list of endangered species in 1967.

RECOVERY AND CONSERVATION. The Red Wolf Recovery Plan prescribed the development of captive-breeding facilities capable of maintaining 300 Red Wolves. The US Fish and Wildlife Service captured the last remaining Red Wolves in 1973 and established a captive-breeding program at a zoo in Tacoma, Washington. After several trial reintroductions, including an unsuccessful experiment in Great Smoky Mountains National Park, Red Wolves were released at the Alligator River National Wildlife Refuge in North Carolina. There, the population grew to over 120 animals in about 17 packs, but it recently declined to about 75 wolves. Encroachment into the area by Coyotes again threatens these wolves, but efforts have been made to sterilize Coyotes to minimize hybridization.

Although many landowners welcome the wolves, some do not, and there have been numerous illegal killings. In an attempt to inform the public about the value of carnivores, a nonprofit organization has taken thousands of people onto the refuge for evening information sessions, which also feature "howling safaris." Participants get to howl at the wolves and are thrilled when there is a response. Unfortunately, Red Wolf howls no longer echo through the forests of East Texas, and no plans exist to reintroduce the species to the state.

FEDERAL DOCUMENTATION:

Listing: 1967. Federal Register 32:4001
Critical Habitat: Not designated
Recovery Plan: US Fish and Wildlife Service.
    1990. Red Wolf recovery/species survival plan.
    Atlanta, GA. 110 pp.

Jaguar, *Panthera onca*

BRIAN R. CHAPMAN

FEDERAL STATUS: Endangered
TEXAS STATUS: Endangered (Extirpated)

Historically, the Jaguar was a common resident in the southern half of Texas. Well-established populations may have extended from the thorny brushlands bordering the Nueces River to the riparian forests along the Guadalupe River. By 1905, Jaguars had become extremely rare in the state; a Jaguar shot near Kingsville in 1948 represents the last documented record in Texas. The species survives in Mexico, and a few inhabit isolated areas in southern New Mexico and Arizona. The Jaguar is maintained on the state list of endangered species to protect animals that might wander across the Rio Grande.

DESCRIPTION. The Jaguar, the largest of the spotted cats in North and South America, resembles the slightly smaller Leopard of sub-Saharan Africa and parts of Asia. The base coloration of most individuals is a tawny yellow, but the pelage of some is tan or reddish brown.

The underside and inner surfaces of the legs are white. The large head is covered with black spots, and the sides are adorned with black rosettes, which often encircle smaller black dots (fig. 8.17). The spot patterns vary among individuals and usually differ on each side of an animal. The relatively short tail is irregularly covered with black spots, which merge toward the end to form rings. About 6 percent of the population in some areas is composed of melanistic (black) individuals, which are commonly called "Black Panthers" (these are not recognized as a separate species).

DISTRIBUTION. The historical distribution of the Jaguar extended from southern Argentina to the

Figure 8.17. Jaguars once roamed widely through South Texas. Photograph by Charles J. Sharp, Creative Commons.

southern United States. The species once occurred from California to Florida, and Jaguar bones found in a Tennessee cave may indicate the extent of their northern distribution. The largest remaining population is in the rainforest of Brazil. During the past 20 years, several Jaguars have been seen repeatedly in the remote mountains of southern Arizona and New Mexico, indicating they might be residents. There have been no recent observations of the cat in Texas.

HABITAT. Jaguars occupy a variety of habitats ranging from deserts and sandy beaches to tropical rainforests. The common denominators for all habitats are considerable plant cover, a supply of freshwater, and an adequate prey base. In Texas, the cats occurred in the thorn scrub of South Texas, dense forests of East Texas, and scattered Live Oak mottes on the Edwards Plateau.

NATURAL HISTORY. Few descriptions exist regarding the natural history of the Jaguar in Texas, so much has to be inferred from studies conducted elsewhere in its range. Like most large carnivores, Jaguars have large home ranges—the home range estimated for one animal in northwestern Mexico was 39 square miles. The ranges of males, which are larger than those of females, often overlap, but females maintain separate ranges.

Jaguars are opportunistic feeders that stalk or ambush prey. Most prey in Mexico are mammals weighing less than 2.5 pounds. Jaguars also take Collared Peccaries, White-tailed Deer, large ground-dwelling birds, and sea turtle eggs.

Mating occurs in December and January. Females give birth to two to four offspring in dens in caves, rocky crevices, or dense thickets. The kittens are born with a thick coat of spotted fur, and their eyes remain closed for several days after birth. They grow rapidly, reaching the size of a domestic cat within about six weeks. Kittens remain with their mother for up to two years before they disperse.

REASONS FOR POPULATION DECLINE. As settlers moved westward in the 1800s, the Jaguar was quickly eliminated. The big cat, considered a threat to both humans and livestock, was shot on sight and its habitats were significantly altered, further depressing survival. Jaguars were also harvested for their spotted coats. When the demand for spotted furs reached its apex in the 1960s, the cats had been eliminated from the United States. During that decade, about 15,000 Jaguar skins were exported annually from Brazil.

RECOVERY AND CONSERVATION. Current conservation efforts focus on educating ranch owners and promoting ecotourism while a recovery plan is being developed. Conservation of the species is important because some consider Jaguars to be an "umbrella species." Such species have broad home ranges and habitat requirements. When an umbrella species is protected, many other species within the same range and habitats are also safeguarded.

A Jaguar Recovery Team with representatives from the United States and Mexico produced a recovery outline for the Jaguar in 2012. The recovery outline serves to guide direct recovery efforts until a full recovery plan can be developed and approved. Responding to a court order, the US Fish and Wildlife Service used information contained within the recovery outline to designate critical habitat for the species.

FEDERAL DOCUMENTATION:
Listing: 1997. Federal Register 62:39147–39157
Critical Habitat: 2014. Federal Register 79:12572–12654
Recovery Plan: Not developed

Gulf Coast Jaguarundi,
Puma yagouaroundi cacomitli

BRIAN R. CHAPMAN
AND MICHAEL E. TEWES

FEDERAL STATUS: Endangered
TEXAS STATUS: Endangered (Extirpated)

The Jaguarundi, the rarest of all cats in the United States, is most likely extirpated from Texas. Jaguarundis once inhabited the Lower Rio Grande Valley, a region in South Texas dominated by thorny brush habitats near the Mexican border.

Figure 8.18. The Jaguarundi is a secretive, rarely seen animal in the Lower Rio Grande Valley, its former range. Photograph by Bodlina, Creative Commons.

Although citizens occasionally report sightings of the elusive species in southwestern Texas, none have been confirmed in the last four decades. The last documented specimens, both road-killed animals, were collected in 1969 and 1986.

DESCRIPTION. Slightly larger than a large domestic cat, the Jaguarundi is often compared to a weasel or otter. The comparisons are natural because the Jaguarundi has short legs, a long, slender body, and a long tail that is about two-thirds the length of the body. The head is small, elongated, and slim (fig. 8.18). Short, rounded ears are widely separated on the head and lack white spots on the back. Lacking bold markings, the short, sleek fur is uniformly colored in one of two color phases, reddish brown or grayish black. Individuals of both color phases can be found in the same litter and in the same region.

DISTRIBUTION. Only one of the seven recognized subspecies, the Gulf Coast Jaguarundi, occupied a small range in Texas. The distributional range of the Gulf Coast Jaguarundi stretched northward in the coastal lowlands along the Gulf of Mexico from Veracruz, Mexico, to the Lower Rio Grande Valley of South Texas. It was once present in Cameron, Hidalgo, Starr, and Willacy Counties in Texas but may have ranged more widely in South Texas.

Jaguarundi sightings are often reported from areas outside their known range. While some of these reports are based on incorrect identifications, sightings in the late 1980s in Brazoria County, Texas, and in various parts of Louisiana were probably of released pets. A small population was established in Florida through introductions by humans and is believed to persist.

HABITAT. Across its range, the Jaguarundi occupies various habitats but prefers habitat mosaics of dense thorny brush interspersed with open grassy areas or low shrub cover. Travel and hunting are usually concentrated along the edges of dense cover. The Lower Rio Grande Valley is typified by dense thickets of cacti and thorny trees and shrubs. The vegetation is similar to that of native brush communities elsewhere in South Texas but is often taller and more botanically diverse, with plant species more typical of adjacent Mexico. Native brush areas have been extensively converted to farmland, rangeland, and urban areas, and more than 95 percent of native brush has been eliminated in the Lower Rio Grande Valley.

NATURAL HISTORY. Little is known about the ecology of Jaguarundis in Texas—most information about their life history is gleaned from studies of the species in Central and South America. There, males have larger home ranges than females. Jaguarundis frequently travel and hunt in pairs and are most active during the day. Jaguarundis sometimes climb trees but forage most often on the ground. Remains of ground-feeding birds, rodents, rabbits, and reptiles are frequently found in Jaguarundi stomachs. Like most felids, Jaguarundis are opportunistic when hunting and occasionally prey on poultry, devour arthropods, and capture fish stranded in puddles.

Little is known about the Jaguarundi mating system. Observations of young in March and August indicate that Jaguarundis in Mexico may have two breeding periods, but in northern Tamaulipas and South Texas, a single, late-autumn breeding period is more common. After mating, the female selects a den site in fallen logs, dense vegetation, or hollow trees. The litters of one to four cubs start leaving the den at 28 days and begin eating by themselves when they are about 42 days old. Females reach sexual maturity when they are 1.6 years old.

REASONS FOR POPULATION DECLINE. Alteration or elimination of thornscrub habitats in Texas is implicated as the major contributing factor to the extirpation of the Jaguarundi from its former range. Some preyed on barnyard poultry or small domestic animals, a behavior that usually ended in death when a landowner trapped or shot the predator. Jaguarundi pelts have no commercial value, but a few die each year at the hands of indiscriminate hunters.

RECOVERY AND CONSERVATION. The US Fish and Wildlife Service recovery plan includes several species of endangered cats in Texas and Arizona. Although the Jaguarundi is mentioned in the recovery plan, its primary focus is the Ocelot. Since the Gulf Coast Jaguarundi and Ocelot occupy the same habitats, and much more is known about the biology and habitat requirements of the latter, an assumption was made that recovery objectives for the Ocelot in Texas would also benefit the potential recovery of the Jaguarundi in Texas and Arizona.

The Texas Department of Transportation is working with the Texas Parks and Wildlife Department to construct bridges, culverts, overpasses, and fences along roadways to prevent Jaguarundi and Ocelot mortality. Critical habitat rules for the Jaguarundi have not been published, but the US Fish and Wildlife Service, the Texas Parks and Wildlife Department, The Nature Conservancy, and many local landowners are acquiring, restoring, and protecting brush habitats in the Lower Rio Grande Valley. Current efforts include planning a habitat corridor between Cameron and Willacy Counties to link previously acquired national wildlife refuge units. In addition, a successful captive-breeding program exists at several zoos. Individuals from these programs may be available for reintroduction once enough habitat has been restored. Until enough habitat

is available to allow reintroduction of captive-bred Jaguarundis, recovery of the species in Texas appears unlikely.

FEDERAL DOCUMENTATION:
Listing: 1976. Federal Register 41:24062–24067
Critical Habitat: Not designated
Recovery Plan: US Fish and Wildlife Service. 2013. Gulf Coast Jaguarundi recovery plan (*Puma yagouaroundi cacomitli*): First revision. Albuquerque, NM. 70 pp.

Ocelot, *Leopardus pardalis albescens*

MICHAEL E. TEWES

FEDERAL STATUS: Endangered
TEXAS STATUS: Endangered

During the first half of the 1800s, Ocelots occurred in South, Central, and East Texas and were especially common in the major river watersheds in southeastern Texas. As settlements expanded in Texas, the Ocelot range contracted and is now confined to two small populations in extreme South Texas.

DESCRIPTION. The Ocelot is a medium-sized cat with heavily spotted fur bearing black rosettes and decorative spots scattered over a tan or yellow background. Coat markings are variable, with different patterns on opposite sides of

Figure 8.19. A trail camera recorded this prowling Ocelot on a trail in Laguna Atascosa National Wildlife Refuge. Photograph courtesy of the US Fish and Wildlife Service.

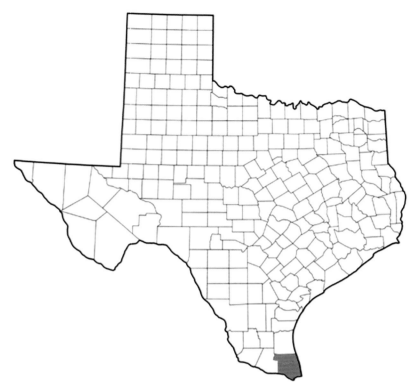

Figure 8.20. The remaining Ocelots currently occupy two widely separated thornscrub habitats in the southern tip of Texas. Map by Brian R. Chapman.

compared to the Cameron situation. Much of the Lower Rio Grande Valley is highly developed with large highways, housing subdivisions, and industrial facilities, especially along the Brownsville Ship Channel, which forms an obstacle for Ocelot migration. The road network and rapidly expanding human population present a significant mortality risk to wandering or dispersing Ocelots.

HABITAT. Ocelots use a variety of plant communities throughout the Western Hemisphere, including primary and secondary forests, tropical forests, mangrove stands, and dry thornscrub. Ocelots exhibit a strong affinity for extremely dense thornscrub communities with closed canopies. The herbaceous or tree layers seem less important than the screening, escape, and ambush cover provided by a dense woody understory. Less than 1 percent of South Texas supports this type of thornscrub, and the remaining habitat patches are small, isolated tracts capable of supporting only a few Ocelots. Most thornscrub tracts lack connectivity. The roads and open habitats separating the tracts hinder movements across the landscape and increase mortality risk. Patch isolation is a particularly significant limiting factor for the Cameron population.

NATURAL HISTORY. The territory of a resident male Ocelot often overlaps the home range of one to three females. Home range size varies but typically encompasses 2 to 3 square miles. Prime habitat often represents a small portion of the entire home range. Resident Ocelots defend their territory to protect access to its resources (e.g., mates, cover, food) from transients or dispersers. A transient may not be able to successfully defend a territory and must constantly move among marginal habitat patches to survive. A disperser moves from the safety of its home range in search of a new territory. The mortality rate of transients and dispersers is usually higher than that of residents.

an individual. The 18-inch tail has black rings (fig. 8.19). Male Ocelots in Texas weigh about 25 pounds, and female Ocelots about 20 pounds.

DISTRIBUTION. The current Ocelot range extends from South Texas to northern Argentina. The original Ocelot subspecies designations were based on pelage and morphometric traits that reflected natural variation. Consequently, the 11 subspecies likely have little taxonomic or ecological utility. The type specimen for *Leopardus pardalis albescens* was described from Louisiana, but the subspecies ranged across Texas to the Rio Grande and beyond.

Of the two remaining populations in the state, the "Cameron population" consists of about 12 to 15 Ocelots inhabiting Laguna Atascosa National Wildlife Refuge in Cameron County. The refuge is isolated by open environments of agricultural land, saline coastal prairies, and numerous roads. A significantly larger "Willacy population" occurs on private ranches in and around northern Willacy County (fig. 8.20). The rangelands are separated by fewer roads between patches of prime habitat

Both Ocelot populations in South Texas experience significant mortality from Ocelot-vehicle collisions. Ocelots also die from disease and rattlesnake bites, and from Ocelots killing other Ocelots. Ocelots are primarily nocturnal, with activity peaks during twilight. Some activity occurs during the day, but during hot summer afternoons Ocelots rest in cool, shaded locations. Behavior during active periods includes territorial patrolling, scent marking, and foraging, particularly when their prey is active.

Ocelot prey consists of small mammals weighing less than 2.2 pounds. Their main prey consists of Eastern Cottontails, rodents, birds, and reptiles. Extended drought causes small mammal populations to decline and likely results in nutritional deficiencies, reduced reproduction, and increased Ocelot mortality.

After a gestation period of 78 to 83 days, a litter of one to three kittens is born. The kittens are altricial—their eyes remain closed for 14 days, and they rely on their mother's care for the first several months of life. Male Ocelots do not assist in raising young. Male Ocelots usually leave their natal home range after reaching adulthood, whereas a female Ocelot may share a portion of its mother's home range or settle nearby.

REASONS FOR POPULATION DECLINE AND THREATS. Lack of prime habitat areas for Ocelots is the fundamental constraint to Ocelot recovery. Population isolation has fostered inbreeding, thereby contributing to genetic erosion, which has negative impacts on reproduction and survival. Roads are the primary proximate agent for Ocelot

mortality. Projections of future increases in the human population and related developments will intensify habitat isolation and threaten population survival, particularly for the Cameron population.

RECOVERY AND CONSERVATION. Although habitat restoration is listed in the recovery plan as a recovery strategy, few efforts have provided significant areas of usable habitat. If the overall techniques for achieving restored habitat can be identified and implemented, costly investments will be required to develop enough significant habitat to benefit Ocelot recovery.

Private landowners control lands that support 80 percent or more of the remaining Ocelot population in Texas. Avoidance of regulations or agency programs that reduce management options on private lands because of Ocelot presence should be discouraged. Instead, incentives from nongovernmental organizations to assist Ocelot recovery on ranches could have a significant positive impact on Ocelot recovery.

FEDERAL DOCUMENTATION:
Listing: 1982. Federal Register 47:31670–31672
Critical Habitat: Not designated
Recovery Plan: US Fish and Wildlife Service.
    2016. Recovery plan for the Ocelot (*Leopardus pardalis*), first revision. Albuquerque, NM.
    217 pp.

9 RESIDENT AND MIGRATORY BIRDS

It was a spring without voices. On the mornings that once throbbed with the dawn chorus of robins, catbirds, doves, jays, wrens and scores of other bird voices, there was now no sound; only silence lay over the fields and woods and marshes. — RACHEL CARSON, 1962

The haunting possibility of a spring morning unaccompanied by the melodies of birds inspired Rachel Carson to write *Silent Spring*, an evocative warning about the effects of pesticides and other chemical pollutants in the environment. Widespread applications of DDT and other organochlorine pesticides had an especially devastating effect on the populations of many avian species, especially those at the top of the food chain. Populations of the Bald Eagle, Peregrine Falcon, Osprey, Brown Pelican, and others suffered steep declines as the hydrocarbon pesticides became concentrated in their body tissues and interfered with the production of normal eggshells. Brown Pelicans in Texas faced extinction because their fragile eggs were crushed when a parent attempted incubation. Fortunately, the efforts of Rachel Carson and a growing environmental movement in the 1960s led to a ban on the use of DDT and related organochlorine pesticides in time to prevent the extermination of many vulnerable species, most of which have substantially rebounded.

Local populations of birds fluctuate from year to year depending on climatic variability and its effect on habitat and food supply. Short-term declines are typically followed by rebounds in healthy populations, but when measured over longer periods, bird populations can serve as analogues to the Canary in the coal mine (chapter one). Birds are sensitive indicators of environmental health. Two national surveys monitor long-term bird population trends and provide insights into ecosystem viability. The Christmas Bird Counts, which began in 1900, document changes in early winter bird populations throughout North America, and the North American Breeding Bird Survey, initiated in 1966, counts birds annually in June and July at prescribed points throughout the nation. The data from these surveys provided an early warning that Neotropical migrants (birds that breed in North America and spend the winter months in Central or South America) were undergoing steep population declines.

Habitat quality and expanse, more than any other factors, determine population size. Many endangered species, including the Red-cockaded Woodpecker, Interior Least Tern, and Southwestern Willow Flycatcher, are habitat specialists with limited distributions where local habitats meet their needs. Migratory species are confronted with a three-pronged habitat problem—habitat quality must be sufficient where they breed, where they overwinter, and where they stop to refuel and rest during migrations. Data from the two nationwide surveys indicate that populations of many grassland and wetland birds are also declining as their habitats are disappearing or becoming increasingly fragmented. Most of the threatened and endangered species described in the accounts that follow have been affected by habitat degradation.

Whooping Crane, *Grus americana*

JULIA C. BUCK

FEDERAL STATUS: Endangered
TEXAS STATUS: Endangered

Although the Whooping Crane is still one of the rarest birds in North America, its recovery from the brink of extinction is lauded as one of the greatest success stories of the conservation movement. The only remaining natural migratory population, the Aransas–Wood Buffalo Population (~300 individuals), overwinters at the Aransas National Wildlife Refuge (ANWR) near Rockport, Texas, and breeds at Wood Buffalo National Park (WBNP) in Canada. Experimental populations of Whooping Cranes include the Eastern Migratory Population (~100 individuals), which overwinters at the Chassahowitzka National Wildlife Refuge in Florida and breeds in the Necedah National Wildlife Refuge in Wisconsin, and two nonmigratory populations, one in Florida (~20 individuals) and one in Louisiana (~24 individuals).

DESCRIPTION. The 4.9-foot-tall Whooping Crane is the tallest bird in North America. It has a wingspan of 7.5 feet and an average weight of about 15 pounds. Males are generally slightly larger than females. This magnificent bird is snowy white with black wing tips. A red crown extends from the cheek, along the bill, and over the top of the head. The legs are black, the bill is dark olive gray, and the eyes are yellow. Juveniles, which have a cinnamon-brown head and lack the red crown, acquire adult plumage late in their second summer (fig. 9.1).

The Whooping Crane is named for its distinctive "contact call," a loud, trumpeting bugle expressing reassurance and location. Other calls warn family group members of potential danger, serve to strengthen pair bonds, and defend territory.

The only other crane in North America is the Sandhill Crane, which is easily recognized by its gray-brown color. Several other birds—Great Egret, Great White Heron, and Wood Stork—are similar in appearance to the Whooping Crane and can be misidentified when viewed from afar. These are significantly smaller than the Whooping Crane and lack the characteristic red crown.

DISTRIBUTION. Historically found from southwestern Louisiana through the Rio Grande

Figure 9.1. Each young Whooping Crane that arrives at Aransas National Wildlife Refuge to spend the winter renews hope for the survival of the species. Photograph by David Rein.

Delta in northeastern Mexico and inland on grasslands including the tablelands of west-central Texas and high plateaus of central Mexico, Whooping Cranes are now most easily observed at ANWR in Texas. Before European settlement, the population was estimated at more than 10,000 birds. By the time ANWR was created in 1937 to protect the remaining birds' winter range, hunting and habitat loss had driven the population to the brink of extinction. Since then, the population has rebounded and its winter range has expanded to include nearby coastal areas, including Matagorda Island and various private lands.

Whooping Cranes complete an annual migration of nearly 2,500 miles from their breeding grounds at WBNP to their wintering grounds at ANWR. The migration route follows the Central Flyway from Alberta to Texas (fig. 9.2). Stopover sites include southern Saskatchewan, the central Platte River of Nebraska, and Quivira National Wildlife Refuge in Kansas. Migration begins in mid-September, and cranes arrive at ANWR in November. The spring migration begins in late March or early April, and cranes arrive at WBNP in late April or early May. Occasionally, a few birds remain on the Texas coast throughout the summer.

HABITAT. On the wintering grounds, Whooping Cranes usually forage in shallow bays, open ponds, salt marshes, and sand flats on the margins of the mainland and barrier islands. Roosting occurs in vegetated marsh habitat dominated by Saltwort, Sea Ox-eye, Glasswort, and Carolina Wolfberry. During drought years when marsh salinities may exceed 20 parts per thousand, Whooping Cranes occasionally move inland to forage in and drink freshwater.

Whooping Cranes use many habitats, but primarily wetland mosaics, for feeding and roosting during migration. They also use riverine habitats, particularly in Nebraska. Whooping Cranes breed in wetland habitats at WBNP. Dominant vegetation in the potholes used for nesting includes bulrushes, cattails, sedges, and Muskgrass.

NATURAL HISTORY. The life span of the Whooping Crane is up to 30 years. The species is

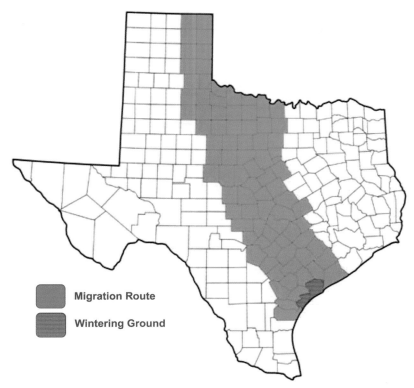

Figure 9.2. Whooping Cranes could be observed anywhere on their migratory route to and from their breeding grounds in Canada. Map by Brian R. Chapman.

omnivorous, but animal material is favored. Blue Crabs constitute up to 90 percent of energy intake on the wintering grounds. Carolina Wolfberry fruits are a particularly important food source when cranes arrive on the wintering grounds in the fall. They also consume mollusks, insects, fishes, frogs, small mammals, acorns, and agricultural grains.

Whooping Cranes reach sexual maturity at three to four years of age and mate for life, although they may select another mate following the death of the first. Elaborate courtship displays are initiated on the wintering grounds and include jumping up and down, leaping, bowing, head bobbing, vocalizations, and wing movements. Pair bonds are reinforced during migration. Upon arrival at the breeding grounds, the pair mates, builds a nest, and lays one to three eggs (usually two) in late April to mid-May. A nest site may be used for several successive years. The incubation period is 29–31 days and parental care is shared, but most nests produce only one surviving offspring. The juvenile completes its first migration in the company of its parents and remains with them for its first winter.

On the wintering grounds, a family group defends a territory from other family groups. Recent population increases have resulted in shrinking territory size on the wintering grounds.

REASONS FOR POPULATION DECLINE AND THREATS. Hunting of Whooping Cranes, which was legal until 1918, was a significant factor in the historical population decline of the species. Documented shootings of Whooping Cranes in the late 1800s and early 1900s numbered in the hundreds, and most mortalities occurred during migration. Despite the potential for severe penalties, Whooping Cranes are still illegally shot, usually during migration, although this is now a rare occurrence.

Habitat degradation and loss also impacted the historical population. Conversion of historical breeding and migratory habitats to agriculture in the northern Great Plains and prairie pothole region of the United States and Canada contributed to the population decline. Habitat loss on the wintering grounds continues to be a concern. More than 50 percent of historical tidal salt marsh habitat on the Texas coast has been lost, and the elimination of habitat because of development and erosion remains a significant threat, especially as the Aransas–Wood Buffalo Population increases.

Approximately 60–80 percent of Whooping Crane mortalities occur during migration, when birds encounter novel hazards. Although cause of death is difficult to establish, collision with utility lines and fences is probably important.

Limitation of freshwater inflows from the Guadalupe and San Antonio Rivers into San Antonio Bay may negatively impact Whooping Cranes. Sufficient inflows are needed to maintain estuarine health and conditions favorable for prey species (especially Blue Crabs and wolfberries) and to ensure availability of fresh drinking water for cranes. However, upstream reservoir construction and diversion of water for municipal, industrial, and agricultural uses is increasing. The Texas Parks and Wildlife Department has recommended target inflow levels to maintain the health of the bay, but mechanisms to ensure water allocation are not mandated by law. In 2010, an environmental group, the Aransas Project, filed a lawsuit against the Texas Commission on Environmental Quality (TCEQ), claiming that its water management practices caused the deaths of nearly two dozen Whooping Cranes in the winter of 2008–2009, thereby violating the Endangered Species Act. The Aransas Project won this suit, and the TCEQ was ordered to stop issuing new water rights and write a habitat conservation plan to ensure adequate flow to the estuary.

Because the Aransas–Wood Buffalo Population constitutes a small, isolated population in a limited geographic area, it is extremely vulnerable to random events such as chemical spills and hurricanes. Limited genetic diversity is also a concern.

RECOVERY AND CONSERVATION. Whooping Crane recovery efforts involve many agencies in the public, private, and nonprofit sectors in the United States and Canada. The Aransas–Wood Buffalo Population has experienced substantial growth (~4.6 percent per year) over the past seven decades. From a low of 15 individuals in 1941, the population has recovered to nearly 300 individuals, and population viability analyses suggest a low probability (< 1.0 percent) of extinction over the next 100 years.

Areas designated as critical habitat include wintering grounds at ANWR and vicinity and migratory stopover habitats at the Cheyenne Bottoms Wildlife Area and Quivira National Wildlife Refuge in Kansas, the Platte River bottoms in Nebraska, and the Salt Plains National Wildlife Refuge in Oklahoma. Potential critical habitat includes areas in and around WBNP and various wetland staging areas in south-central Saskatchewan.

The stated goals of the current recovery plan are to (1) protect and enhance breeding, migratory, and wintering habitats of the Aransas–Wood Buffalo Population; (2) reintroduce and establish additional self-sustaining wild populations within the species' historical range that are geographically separate from the Aransas–Wood Buffalo Population; and (3) maintain a captive breeding population to protect against extinction. Because considerable time is necessary to reach downlisting goals, delisting criteria have not been established.

FEDERAL DOCUMENTATION:

Listing: 1967. Federal Register 32:4001

Critical Habitat: 1978. Federal Register 43:20938–20942

Recovery Plan: Canadian Wildlife Service and US Fish and Wildlife Service. 2007. International recovery plan for the Whooping Crane. Albuquerque, NM. 162 pp.

Eskimo Curlew, *Numenius borealis*

MARY KAY SKORRUPA

FEDERAL STATUS: Endangered (Likely Extinct)
TEXAS STATUS: Endangered

The Eskimo Curlew was one of the most abundant shorebirds in North America, but then its numbers plunged to so few that the species was believed to be extinct as early as 1905. In 1833, flocks of Eskimo Curlews were so dense in Labrador that they reminded John James Audubon of Passenger Pigeons. Likewise, flocks during spring migration in the plains were reported to cover 40–50 acres when they landed. Despite the rapid demise of their numbers around the beginning of the twentieth century, sightings of the species continued for many decades thereafter. In 1962, a small flock was documented in Texas, and another was seen in 1981, but there have been no accepted observations of the species anywhere since 1987. Although unconfirmed sightings are reported every few years, the Eskimo Curlew is likely extinct.

DESCRIPTION. The long legs and long, slightly downcurved bills of all curlews give them a distinctive appearance. Although smaller (12 inches in length), the Eskimo Curlew was similar in appearance to the Hudsonian Curlew, a subspecies of Whimbrel (fig. 9.3). As in most curlews, the upperparts of the Eskimo Curlew were mottled brown, but the underparts were lighter brown. A pale stripe bisected the dark crown on the head. The distinguishing features were best seen on a flying Eskimo Curlew—the undersides of the wings were a rich cinnamon and the primary feathers lacked barring.

Figure 9.3. This painting by Louis Agassiz Fuertes (1874–1927) illustrates the size of the Eskimo Curlew (center) in relation to the Long-billed Curlew (top left), Hudsonian Curlew (bottom left), Marbled Godwit (top right), and Hudsonian Godwit (bottom right). From Wikimedia Commons.

DISTRIBUTION. The breeding grounds were on the treeless Arctic tundra in western Canada and possibly in Alaska. Huge flocks migrated eastward across Hudson Bay to coastal Labrador and Maine after the breeding season. Before migrating southward to Patagonia and Argentina, the birds fattened themselves on Crowberries, a habit that gave rise to the name "Doughbird." The return migration was along the Pacific Coast, through Central America, and across the Gulf of Mexico to the Texas coast. After a short stay in the estuarine and prairie habitats in Texas, migration continued northward through the High Plains to the Arctic breeding grounds.

HABITAT. Nests were in Arctic tundra habitats dominated by grasses and sedges. Scattered clumps of stunted birch were also present in some portions of the breeding range. During fall migration, flocks of Eskimo Curlews preferred heath habitats in northern Canada, pastures in the Great Plains, and intertidal flats along the Gulf of Mexico. In South America, the birds may have preferred intertidal zones, wet pampas grasslands, and semidesert zones, but their habitat preferences while in South America are poorly known. While migrating northward through North America in the spring, the curlew visited estuarine habitats along

the Texas
and mixe
favored g
the grass

NATUR
in an ina
breeding
were scr
soil—co
clutch ty
brown s
and cap
hatching.

Eskimo Curlews fed by probing the soil for invertebrates and picking insects and berries from vegetation. During the spring migration through the High Plains of North America, they preyed on the eggs and adults of the Rocky Mountain Grasshopper. Their food habits while in Texas are unknown, but one observer in 1884 noted that they seemed to seek the same food as the American Golden-Plover.

REASONS FOR POPULATION DECLINE. The primary cause of the Eskimo Curlew population decline was unregulated hunting; thousands of the birds were killed each year during migration through the High Plains. Some were shot for sport, but most were shipped to eastern markets for table fare. Their large flocks sometimes covered tens of acres, and their apparent lack of fear made them easy targets for market hunters. During the same period, their migratory habitats in both North and South America were being converted to agriculture and fires were being suppressed, which allowed brushy vegetation to invade shortgrass prairies. The extinction of the Eskimo Curlew's principal food source during spring migrations, the Rocky Mountain Grasshopper, likely contributed to the plight of the species.

RECOVERY AND CONSERVATION. Development of a plan to conserve the Eskimo Curlew awaits rediscovery of a viable population.

FEDERAL DOCUMENTATION:
Listing: 1967. Federal Register 32:4001
Critical Habitat: Not designated
Recovery Plan: Not developed

Western Atlantic Red Knot,
Calidris canutus rufa

KIM WITHERS

FEDERAL STATUS: Threatened
TEXAS STATUS: Not Listed

The Red Knot is a cosmopolitan shorebird species well known for its long-distance migrations. The Western Atlantic Red Knot travels nearly 10,000 miles one way from Arctic breeding grounds to wintering grounds in Tierra del Fuego at the tip of South America. The Texas Gulf Coast is an important wintering and migratory stopover area for the Western Atlantic Red Knot. Populations of Red Knots are declining worldwide, but the *rufa* subspecies has declined by 67–88 percent since the 1980s to an estimated 18,000–35,000 in 2008.

DESCRIPTION. Although Red Knots are the largest sandpipers in the Western Hemisphere, they are medium-sized shorebirds with a typical "sandpiper" profile: small head, short neck, small eyes, a relatively short and straight bill, and fairly stout legs with the lower leg longer than the upper leg. Legs and bills are black. The breeding plumage is characterized by white lower flanks, underside, and undertail coverts; a soft, chestnut-colored belly; a grayish nape; and a silvery gray back. Males in breeding plumage are usually more brightly colored and have a reddish eye line, whereas females are duller and have an indistinct eye line

(fig. 9.4, left). Their winter plumage is dull gray dorsally with white underparts—the sexes are indistinguishable (fig. 9.4, right).

DISTRIBUTION. The *rufa* subspecies of Red Knot breeds in the central Canadian Arctic (north of 60°) from the west coast of Baffin Island west to Victoria Island and the islands of northern Hudson Bay. The subspecies migrates through and winters in parts of the United States, the Caribbean, and South America, with well-known spring and fall migratory stopovers along the US Atlantic Coast. For example, in the spring more than 50 percent of the population usually stops to feed on Horseshoe Crab eggs along Delaware Bay. Coastal Massachusetts was an important staging area for spring migrants in the late 1800s, but few stop there today.

HABITAT. In Texas and other coastal areas, Red Knots occupy intertidal shoreline habitats including tidally influenced sandy beaches and mudflats. Mangroves, brackish lagoons, salt marshes, and tidal creeks may also be used. They rarely wade, preferring exposed, saturated areas. Both staging and wintering habitats are typically characterized by high-energy waves and/or currents where food quality and availability are maximized.

NATURAL HISTORY. After sharing some staging areas during their fall migration, the population of the *rufa* subspecies separates to winter in Tierra

Figure 9.4. Red Knots discard their namesake summer plumage (left) for a less colorful winter plumage (right). Photographs by Brent Ortego (left) and Elizabeth Smith (right) were taken on Padre Island, Texas.

del Fuego, northern Brazil, the US Atlantic and Gulf Coasts, and coastal habitats of the Caribbean Sea. Some *rufa* Red Knots spend nearly 80 percent of their annual cycle in the coastal habitats of Texas and northern Mexico, from Matagorda Island south to Tamaulipas. Birds wintering in Texas depart for the breeding grounds in mid to late May and return in late July to mid-August; time on the breeding grounds is 70–95 days. The *rufa* Red Knots use the midcontinental flyway on both northbound and southbound migrations. Many southbound Red Knots spend one to two weeks at a few (mostly Canadian) stopovers such as wetlands in southern Hudson Bay and Saskatchewan before embarking on the final leg—a two- to three-day nonstop flight to the Gulf Coast. Birds that continue to more southerly wintering areas arrive in South America from mid to late August through September.

Depending on snow cover, Red Knots arriving on coastal tundra in Arctic breeding areas disperse to nesting habitats one to two days after arriving. Males generally arrive before females, establish territories, and begin song flights; pairs form within a few days of the arrival of the females. Males construct three to five nest scrapes—cup-shaped depressions lined with dried leaves, grasses, and lichens—in their territories on southwest-facing, raised rocky areas with little vegetation. Only one brood is raised each season, but a replacement clutch may be laid if the first clutch is lost early. Clutches usually consist of four eggs laid over six days. Incubation begins when the second egg is laid and continues for about three weeks after the last egg is laid. Both males and females incubate. All eggs in a clutch hatch within 24 hours, and chicks begin moving from the nest to nearby wetland habitats almost immediately. Females abandon the brood within a few days of hatching, leaving the males to care for the young. Males brood chicks until they are about 10 days old, leading them to feeding habitats and defending them from predators until they fledge, about 18 days after hatching.

In wintering areas, Red Knots feed primarily on small mollusks, especially bivalves. Coquina clams are the main prey of foraging Red Knots on Texas beaches; other bivalve prey includes mussels, surf clams, tellins, and macomas. Small gastropods, crustaceans, and polychaetes are also eaten. Red Knots forage by surface pecking, deep probing (often using the full length of the bill), and "plowing," which involves inserting the tip of the bill into the substrate and rapidly moving the head up and down while moving forward. Sensitive nerve endings called Herbst corpuscles in the tip of the bill facilitate prey detection in wet substrates.

REASONS FOR POPULATION DECLINE AND THREATS. Red Knot populations, like those of other Arctic-breeding sandpipers, are most affected by adult survivorship and recruitment, food supplies outside the breeding grounds, and snow depth during incubation. Habitat losses have reduced their resilience, and other factors, such as Horseshoe Crab harvesting to obtain their blue, copper-rich blood for medical applications, have decreased food availability at critical times in the Red Knot's annual cycle. During spring migration, a large proportion of *rufa* Red Knots is particularly dependent on Horseshoe Crab eggs in Delaware Bay. Northbound birds leaving stopovers like Delaware Bay without gaining sufficient weight have lower survival rates—departure weights and reproductive success are linked. Because *rufa* Red Knots tend to concentrate in a few key stopover and wintering sites containing specific food resources, they are particularly vulnerable to environmental changes.

Climate change may cause alteration of habitat as well as timing mismatches between food availability during migration and chick growth. Warming in the Arctic is making tundra breeding habitat shrubbier. If the weather there becomes more unpredictable, the Red Knot's already low fecundity may decline further, causing breeding productivity to become more erratic. Availability of insects for both chicks and adults may be reduced if warming causes insects to hatch earlier. Climate change may also cause asynchronies between migration timing and availability of food resources at migration stopovers.

Fortunately, harvest quotas for Horseshoe Crabs have been reduced in the hope of ensuring enough eggs for migrating Red Knots. However, increases in water temperature may cause Horseshoe Crabs to spawn earlier than normal; severe storms may also delay spawning. In either scenario, Red Knots would not be able to gain sufficient energy reserves to complete migration and reproduce. In addition, relative sea-level rise will likely inundate temperate coastal habitats used by migrating and wintering birds, which will also impact food availability in staging and wintering areas. Shoreline stabilization and hardening in response to sea-level rise will degrade habitat and prevent re-formation of intertidal habitats at higher elevations.

Coastal development and shoreline stabilization have decreased Red Knot habitat throughout its range in the Western Hemisphere. About a third of the remaining coastal habitat in the United States is at risk of development, and significant wintering and stopover habitats from Canada to South America are also at risk. Beach recreation and the disturbance caused by people and their pets, sand placement projects, off-road vehicles, and beach maintenance may all affect the quality of Red Knot beach habitat. While hunting no longer threatens Red Knots in the United States, they continue to be legally and illegally killed for food and sport in South America and the Caribbean.

RECOVERY AND CONSERVATION. The US Fish and Wildlife Service has not developed a recovery plan for *rufa* Red Knots. The goal of *rufa* Red Knot conservation stated in the Western Hemisphere Shorebird Reserve Network conservation plan (2010) is to increase the overall population to between 100,000 and 150,000 birds, the size of the population in the late 1970s. This population size might be achieved by (1) recovering Horseshoe Crab egg densities in Delaware Bay so that the 80,000 Red Knots and all other shorebirds that use the resource can be supported; (2) reducing human disturbance at key stopover sites; (3) protecting all key stopover and wintering sites in the Western Hemisphere that support at least 100 Red Knots; and (4) protecting significant breeding areas. The conservation plan identifies the Delaware Bay shore in New Jersey as a critical spring stopover habitat.

After *rufa* Red Knots were listed in 2014, efforts to protect habitat increased in both the United States and South America. Several key wintering areas in South America are designated to be protected as shorebird reserves, and hunting restrictions are being implemented (Red Knots were recently protected from hunting in French Guiana). In the United States, actions range from beach management that reduces disturbance to a multispecies adaptive resource management framework implemented for Delaware Bay to allow continued Horseshoe Crab harvest, but with assurance that enough eggs will be available for Red Knots and other shorebird species. The US Fish and Wildlife Service will issue a proposal to designate *rufa* Red Knot critical habitat once it determines specific areas that are essential to its conservation; a recovery plan is being developed.

FEDERAL DOCUMENTATION:
Listing: 2014. Federal Register 79:73706–73748
Critical Habitat: Not designated
Recovery Plan: Not developed

Piping Plover, *Charadrius melodus*

KIM WITHERS

FEDERAL STATUS: Threatened (Great Plains Populations)
TEXAS STATUS: Threatened

The Piping Plover is named for its musical call. This endangered shorebird, with a population of about 5,000 breeding pairs, winters on the beaches and bayside tidal flats of the Texas Gulf Coast. The Texas coast is the most important wintering area for the species, particularly for members of the Great Plains population that breed in the northern Great Plains and Canadian prairie provinces. Up to 75 percent of the US population and an average of 52 percent of all Piping Plovers winter in Texas. The juxtaposition of barrier island beaches, bayside sand, mud and algal flats, and tidal passes provides birds with the habitat mosaic necessary to meet their winter foraging and roosting needs regardless of tidal cycle.

DESCRIPTION. Piping Plovers are small, sand-colored shorebirds with white underparts, bright orange legs, and an orange bill with a black tip. Birds in breeding plumage have a single black

Figure 9.5. Piping Plover in winter plumage on Padre Island, Texas. Photograph by Mdf, Creative Commons.

breast band and a black bar across the forehead. During winter, the black chest band is broken or missing, the black forehead band is lost or less pronounced, the legs fade to lighter orange or pale yellow, and the bill becomes mostly black (fig. 9.5). Males and females are often indistinguishable, especially in nonbreeding plumage; in breeding plumage, the black bands on females may be more restricted and duller black.

DISTRIBUTION. The breeding range extends from Nebraska through the northern Great Plains (including the western Great Lakes) into southern Canada (western Alberta, Saskatchewan, and Manitoba). The species also nests on beaches along the northern Atlantic coastline from North Carolina to eastern Canada and the French islands of Saint Pierre and Miquelon. Piping Plovers winter on the southern Atlantic and Gulf of Mexico coasts from the Carolinas to the Yucatán Peninsula. The winter range also includes the Bahamas and the West Indies.

HABITAT. Habitat used in coastal Texas includes barrier island beaches, tidal passes, and unvegetated bayside flats along the entire Gulf Coast. Expansive wind-tidal flats on the Laguna Madre shoreline of Padre and Mustang Islands provide wintering birds with extensive foraging habitat. Other significant wintering areas include shoreline habitats on Matagorda Island, Bolivar Flats, and San Luis Pass on Galveston Island. While

migrating, Piping Plovers briefly occupy lake or reservoir shorelines.

NATURAL HISTORY. Piping Plovers are short-distance, narrow-band migrants. The population wintering in Texas migrates 2,000–3,000 miles (one way, from ~25° N to ~55° N) along a path confined mostly to the area between 90° W and 100–115° W. These birds usually migrate nonstop each season and rarely use migratory stopovers. Northbound migrants move from bayside tidal flats to beach staging areas where they sometimes gather in large flocks; migration peaks in mid-April, and most wintering birds leave Texas by mid-May. Adults return to Texas beaches from mid-July to September; juveniles may continue to arrive in small numbers through November.

Piping Plovers spend 8–10 months on the wintering grounds in Texas, where the availability of winter food resources plays a large role in their overall mortality. They spend much of their time foraging on estuarine invertebrates, both infaunal (living in the substrate) and epifaunal (living on the substrate), including polychaetes, amphipods, and insect larvae; various flies associated with estuarine habitats may also be taken in flight.

Piping Plovers are sight foragers, preferring habitats covered with only a film of water. Their foraging behavior is described as "run and peck," but because their activities are so rapid, they appear to peck randomly. Amid a film of water, such as after a wave, they vibrate an extended foot against the saturated substrate, perhaps to bring prey to the surface. Piping Plovers also forage in drier areas of shorelines, especially along wrack lines where surface-dwelling insects, semiterrestrial amphipods, and spiders may be found.

Early in the season (July–August) when bay-wide water levels are low and temperatures are high, plovers congregate on beaches, where prey is more plentiful than on exposed shoreline flats. Rising water levels in the bays through September and October spur invertebrate recruitment in the shoreline flats. When water levels begin to fall in November, the newly "seeded" flats are exposed and attract the birds where prey is plentiful. These habitats also provide shelter from weather systems that bring strong winds, rough waves, colder temperatures, and sometimes rain between December and February. Piping Plovers often roost in the relative safety of tidal passes.

REASONS FOR POPULATION DECLINE AND THREATS. Recreational, residential, and commercial development of sandy beaches and other shorelines has reduced nesting and wintering habitat. Human activities disturb nesting and wintering Piping Plovers. Beach maintenance reduces prey densities and alters nutrient flows by removing driftwood or other debris that also protects nests or provides roosting cover.

Modifications of river flows, oil and gas development, agricultural development, invasive species and vegetation growth, and density-dependent intraspecific aggression are the greatest threats to Piping Plovers on the breeding grounds. Damming and channelizing of large rivers of the Great Plains, such as the Platte and Missouri, have eliminated extensive areas of riverine sandbar nesting habitat.

Piping Plovers exhibit fidelity to a specific wintering area, which often encompasses several sites close to one another. This characteristic renders the plovers vulnerable to loss and degradation of wintering habitat. Accelerating sea-level rise and loss or modification of winter habitat from development, dredging, shoreline armoring, land subsidence, and invasive vegetation are key threats. Since the 1950s, the area occupied by tidal flats on the central Texas coast has declined by more than 50 percent; many tidal flats are now covered by marshes, seagrasses, or open water. The proximity of many tidal flats to the Gulf Intracoastal Waterway renders them susceptible to petrochemical spills or pollution from other hazardous materials.

RECOVERY AND CONSERVATION. The recovery objective for the northern Great Plains Piping Plover population is to restore it so that its likelihood of extinction is less than 5 percent between 2035 and 2085. The recovery strategy focuses on restoring ecosystem function in breeding and wintering habitats.

Piping Plovers exhibit strong intra-annual and interannual wintering site fidelity, which reinforces the need to maintain quality habitat throughout the species' range. Because the plovers depend on a mosaic of coastal habitats, a juxtaposition of bay and beach foraging and roosting habitats is necessary to consistently meet the needs of the wintering population. Recovery criteria include maintaining availability of winter habitats related to food acquisition, shelter, and roosting. These habitats include intertidal areas on ocean beaches, flats on bays, associated dune systems and flats above annual high tide, and seasonally emergent sandbars, mudflats, and oyster reefs. Critical habitat containing these elements was designated in 2001 based on use patterns at the time, but because coastal ecosystems are dynamic and relative sea level is rising, bird distribution and habitat use have shifted slightly, and additional areas have been identified as conservation priorities. Ensuring that sufficient suitable, functioning habitats are available for Piping Plovers will also safeguard other shorebird species that use the same or similar habitats.

FEDERAL DOCUMENTATION:

Listing: 1985. Federal Register 50:50726–50734

Critical Habitat: 2001. Federal Register 66:36038–36143

Recovery Plan: US Fish and Wildlife Service. 2015. Recovery plan for the northern Great Plains Piping Plover (*Charadrius melodus*) in two volumes. Volume 1: Draft breeding and recovery plan for the northern Great Plains Piping Plover (*Charadrius melodus*). 132 pp. Volume 2: Draft revised recovery plan for the wintering range of the northern Great Plains Piping Plover (*Charadrius melodus*) and comprehensive conservation strategy for the Piping Plover in its coastal migration and wintering range in the continental United States. Denver, CO. 166 pp.

Interior Least Tern,
Sternula antillarum athalassos

ANDREW C. KASNER

FEDERAL STATUS: Endangered
TEXAS STATUS: Endangered

As its common name implies, the Least Tern is the smallest of the North American terns. The three subspecies share identical physical characteristics but are distinguished based on the geographic separation of their breeding ranges. Two of the subspecies nest in Texas: the Coastal (or Eastern) Least Tern (*S. a. antillarum*) breeds along the Gulf Coast, and the Interior Least Tern (*S. a. athalassos*) nests in riverine habitats farther inland. The Interior Least Tern and the California Least Tern (*S. a. browni*), which breeds along the Pacific Coast, are listed as endangered.

DESCRIPTION. Like many terns, the Least Tern appears white when flying because its underparts are white. Its back and the top surfaces of its wings and forked tail are gray. The tern is 9–10 inches long, and its narrow, pointed wings create a wingspan of approximately 20 inches. Its sulfur-yellow bill with a black tip distinguishes the species from other small terns in Texas. During the breeding season, a triangular white patch on the forehead emphasizes the black cap on the head

(fig. 9.6, left). Black eye lines extending from the cap to the bill frame the white forehead patch. Later in the year, the black cap recedes to the back of the head, but the black eye line remains. The yellow legs become brownish gray to black in the fall and winter. Adult males and females are indistinguishable. The upper surfaces of juveniles are brownish gray, shifting to gray in first-year plumage. The cap, nape, and eye lines are grayish, and the eye lines do not extend completely to the bill.

DISTRIBUTION. The breeding range of the Interior Least Tern extends from eastern New Mexico and Colorado to southern Indiana and from Montana to Texas. Traditional nesting locations in Texas include portions of the Canadian River and the Prairie Dog Town Fork of the Red River in the Panhandle. It also nests at sites on the Red River along the Texas-Oklahoma boundary and at two reservoirs, Falcon Lake and Lake Amistad, on the Rio Grande. Historically, it likely nested along the Trinity and Brazos Rivers, where it occasionally occurs in small numbers. The distribution of the Interior Least Tern expanded over the last few decades into the central and western portions of Texas, where it nests on reservoir shorelines exposed by drawdowns. The propensity for the tern to change breeding locations as habitat conditions fluctuate results in interannual variation in colonies across the state.

Figure 9.6. The triangular white patch is obvious on this adult Interior Least Tern as its chick peeks out from the shallow scrape serving as a nest (left). Chicks often wander near the nest until they are ready to fly (right). Photographs by Andrew C. Kasner.

All Least Tern subspecies migrate seasonally to coastal areas in Central and South America during winter.

HABITAT. Interior Least Terns usually nest in small colonies on bare or sparsely vegetated sandbars, islands, or gravel substrates associated with rivers. Most nest locations afford unobstructed views far from bushes or trees that would conceal a predator. Nests are shallow scrapes, often situated near driftwood or other debris that offers concealment of eggs and chicks. Colonies are usually located close to the shallow waters of rivers, lakes, or ponds containing abundant small fishes. In areas where habitat alterations have reduced natural nesting habitats, the Interior Least Tern also nests on alternative sites, such as salt flats, flat gravel roofs of buildings, sand and gravel pits, exposed reservoir shorelines, and newly reclaimed areas of surface mines.

NATURAL HISTORY. Courtship activities begin in early spring (late April–early May) after the terns arrive at a suitable nesting location. Pair formation involves posturing and aerial displays, including aerobatics, noisy pursuits, and "fish flights," in which a male chases a female while carrying a fish in his beak. A male will often perch on buoys, pilings, and piers while holding a fish in his bill and watching for a mate. In the later stages of pair formation, copulation often occurs after the male passes a small fish to the female. Least Terns scrape a shallow depression in a sand or gravel substrate to form a nest. Nearby, small stones, mollusk shells, twigs, or bits of flotsam camouflage the nest site. Like many terns, the Least Tern nests in colonies where nests are spaced 10 feet or more from the nearest neighbor. Piping Plovers, another endangered species, and Snowy Plovers, a species of significant conservation concern, often nest in close association with Interior Least Terns.

During a three- to five-day period in mid to late May, females lay two to three speckled eggs in the scrape. Both parents share incubation duties for 20 to 22 days prior to hatching. Chicks remain in the nest for about a week before venturing out to find shade under vegetation or debris clumps. Although juveniles begin to fly at three weeks of age, their parents feed the chicks until the onset of fall migration (fig. 9.6, right). Juvenile terns often perch on buoys and pilings to await the arrival of a parent with fish.

Throughout the summer, the terns remain close to the nesting colony. Adults hover above shallow waters near the colony before diving headlong to capture small fish or aquatic crustaceans. During the nesting season, Least Terns defend the territory surrounding their nest site from intruders. After giving an alarm call, the birds dive repeatedly at a trespasser and sometimes defecate on a persistent threat. Other members of the colony also participate in defensive activities, sometimes mobbing the intruder.

Fall migration begins in late August. Along the migratory route, Interior Least Terns congregate at staging areas to rest and feed. Most staging areas are on sand or gravel spits at the mouths of streams or rivers or in floodplain wetlands, but they also migrate along the Texas Gulf Coast, where they intermix with Coastal Least Terns. After spending the winter in the Central and South American tropics, Interior Least Terns often return to nesting areas they occupied the previous season.

REASONS FOR POPULATION DECLINE AND THREATS. Before humans began altering the flow of rivers by channelization and dam construction, the summer flow patterns of most rivers in Texas and the Midwest were fairly predictable. Interior Least Terns usually located nests on river sandbars where only unusual and extreme flood conditions could eliminate them. Scouring during spring floods prior to tern arrival removed or prevented encroachment by dense vegetation and maintained suitable habitat conditions. Reservoir discharges alter the historical flow pattern of rivers, adding an element of unpredictability to nesting success. Untimely releases of water—usually cold water from the lake bottoms—often flood colonies during the nesting season and potentially alter fish prey availability. Sediments trapped behind dams starve sandbar development downstream, and reductions in flow allow vegetation to encroach on existing sandbars, reducing the quality and extent of nesting habitat. Pollution entering rivers and streams from industries or urban areas often reduces water quality and fish populations. Human disturbances—especially camping and ATV activities—on riverine sandbars also pose significant threats, frequently causing nest abandonment.

RECOVERY AND CONSERVATION. Although it is difficult to accurately census Interior Least Terns, annual surveys determine the population in each state and monitor nesting success. In some areas, vegetation removal from sandbars enhances potential nesting habitat, and timed water releases from reservoirs prevent colony destruction. Riprap structures placed in rivers divert flow and create new sandbars suitable for tern and plover nests. Efforts to educate recreational users about the presence of and potential harm to nesting birds have been effective in some areas. Recovery efforts have been successful, owing in part to the adaptability of the species to changing habitat conditions and on-the-ground protection efforts in areas where higher densities occur. The five-year review of recovery efforts recommended that the Interior Least Tern be delisted.

FEDERAL DOCUMENTATION:
Listing: 1985. Federal Register 50:21784–21792
Critical Habitat: Not designated
Recovery Plan: US Fish and Wildlife Service. 1990. Recovery plan for the interior population of the Least Tern (*Sterna antillarum*). Twin Cities, MN. 90 pp.

Prairie-Chickens

KIM WITHERS

Attwater's Prairie-Chicken, *Tympanuchus cupido attwateri*
Lesser Prairie-Chicken, *Tympanuchus pallidicinctus*

The booming, whooping, and gobbling calls of prairie-chickens, also known as pinnated grouse, once rang out over vast areas of the North American Great Plains during spring and summer. Males gathered on the "booming" or display grounds (leks) in spring and summer to compete for the right to breed. Their bold behavior rendered them easy targets for hunters and predators, and as the prairie grasslands were put to the plow, converted to pasture, or succumbed to urban sprawl, the birds' population declines accelerated. Prairie-chickens now occupy only portions of their historical range, and their numbers have declined by more than 90 percent in the last 40 years (since the 1970s). Their historical range has contracted by more than 80 percent.

DESCRIPTION. Members of genus *Tympanuchus*, which also includes the Sharp-tailed Grouse, are similar in morphology, plumage, and behavior and are collectively known as prairie grouse. Prairie-chickens are medium-sized ("football-sized"), grayish-brown grouse bearing bands of black, cinnamon, or dark brown alternating with white or buff-colored bands; their underparts are lighter. Females are slightly smaller than males and have similar plumage. Their tails are short and rounded, as are the wings. Both sexes have pinnae (ear-like tufts of elongated feathers) on each side of the neck. Males hold the pinnae erect during courtship displays. Males also have a bright yellow or orange eye comb and red to orange esophageal air sacs on the sides of the neck, which they inflate during courtship.

DISTRIBUTION. The three prairie grouse species were historically distributed from Alaska and the Northwest Territories (Sharp-tailed Grouse only) through the Canadian prairie provinces, Great Plains, and Midwest south to the Gulf Coast of Texas. The extinct Heath Hen, usually classified as a subspecies of Greater Prairie-Chicken, was found along the Atlantic Coast from Massachusetts to Virginia. Extant prairie-chickens are essentially nonmigratory and generally engage in only short-range seasonal or dispersal movements.

HABITAT. Prairie-chickens evolved in extensive landscapes consisting of open, uninterrupted grasslands with scattered trees and shrubs. Breeding habitats include booming grounds with nearby nesting and brood-rearing habitat. Booming grounds are located on large patches of short or sparse vegetation, which allows females to have a clear view of displaying males and all birds to see approaching predators. Nesting and brooding usually occur in areas of greater vegetation cover within 2 miles of the booming grounds. Ground-level cover in brood-rearing areas must provide adequate concealment and protection from weather while allowing easy movement of chicks. Prairie-chickens use the same grassland habitats during fall and winter but extend their foraging activity into nearby fallow grain fields wherever the habitats intermix. Agricultural lands are important for foraging when temperatures are low and insects are less available.

NATURAL HISTORY. Prairie-chickens are sedentary, rarely traveling more than 18 miles. Most individuals remain near the booming grounds all year. Leks are active from March through May. To advertise their fitness at these sites, males establish and defend small territories with elaborate displays in which they extend their eye combs, elevate their tails, droop their wings, erect their pinnae, and extend their necks while dancing forward, shaking, and stamping their feet (fig. 9.7). Expansion of their esophageal air sacs during their dances produces a booming sound. Males may fight with one another when females are present, and serious injury or death can occur. Females also engage in dominance behaviors, such as chasing or flutter dancing and booming. Both males and females are physiologically capable of breeding in their first year, but while females usually do breed, yearling males do not.

Figure 9.7. A male Attwater's Prairie Chicken with pinnae raised is defending his territory on the lek. Photograph by George Lavendowski, US Fish and Wildlife Service.

Nests are shallow, bowl-shaped depressions on the ground lined with dried grasses, leaves, and feathers. Females lay 1 egg per day, producing an average of 11 eggs. Incubation begins after the last egg is laid and lasts 23–26 days. The cryptically colored chicks leave the nest within 24 hours of hatching. Chicks begin making short flights when they are two weeks old, and stronger flights by three weeks of age. Juveniles often flock together or form mixed flocks with adults during winter. Life spans range from 1.6 years for the Greater and Attwater's Prairie-Chickens up to 5 years for the Lesser Prairie-Chicken.

Adults forage on the ground for insects, seeds, leaves, buds, and cultivated grains. In winter, leaves and seeds, especially acorns, predominate in native habitats. During the breeding season and summer, animal matter is a significant diet component. Chicks and juveniles eat insects almost exclusively.

REASONS FOR POPULATION DECLINE AND THREATS. Prior to settlement, prairie grasslands and shrublands covered more than half the North American continent and supported millions of prairie grouse. Unregulated harvest prior to the twentieth century greatly reduced prairie grouse populations throughout their range. Loss, degradation, and fragmentation of habitats caused by conversion to cropland, long-term fire suppression with concomitant increases in trees, excessive grazing and poor grazing practices, herbicide use, energy development (including wind energy), and construction of buildings, fences, and utility lines contributed to additional population declines. Other threats include mortality from collisions with fences, power lines, and cars, and predation, especially where patches of suitable habitat are small and isolated.

The effects of climate change, especially the increased severity and persistence of droughts and extreme weather events, also pose serious threats to the viability of prairie-chicken populations. Prolonged droughts in the Dust Bowl years nearly extirpated Lesser Prairie-Chickens from Colorado, Kansas, and New Mexico and reduced the Texas population. The Lesser Prairie-Chicken still inhabits the five states where it historically occurred, but its range has been reduced by about 83 percent. A series of droughts and floods from 1980 to 1994 devastated the remaining population of Attwater's Prairie-Chicken in Texas, reducing it to fewer than 160 individuals.

Habitat fragmentation has reduced connectivity between prairie-chicken subpopulations. The lek mating system, low nesting success, and increasing isolation have greatly reduced genetic interchange and genetic diversity. Maintenance of genetic diversity is unlikely to be attained because of widespread and ongoing habitat loss.

RECOVERY AND CONSERVATION. The Western Association of Fish and Wildlife Agencies developed a range-wide conservation plan for the Lesser Prairie-Chicken in 2013. The plan, subsequently endorsed by the US Fish and Wildlife Service, emphasizes research on the ecological requirements of the birds and development of conservation agreements by private, public, and industrial landowners. Goals of the conservation plan include identifying core areas containing nesting and brood-rearing habitats, and eliminating threats and stressors including conversion of native grassland to cropland, long-term fire suppression, tree invasion, grazing practices that reduce habitat quality, and habitat fragmentation. Private companies (mostly energy-related companies) in Colorado, Kansas, Oklahoma, New Mexico, and Texas have enrolled more than 4 million acres of prairie habitat under conservation agreements. Aerial surveys in 2015 indicated that the range-wide Lesser Prairie-Chicken population had increased by about 25 percent from 2014 to 2015. The Nature Conservancy assisted the Texas Parks and Wildlife Department in acquiring Yoakum Dunes, a 14,037-acre tract near Lubbock in the Texas Panhandle (Cochran, Terry, and Yoakum Counties). The newly created Yoakum Dunes Wildlife Management Area provides breeding and nesting habitat for the species in an important portion of its range.

Attwater Prairie Chicken National Wildlife Refuge (APCNWR) was established near Sealy, Texas, in 1972, five years after the species was listed,

but more than 10 years before the first recovery plan was completed. Three Attwater's Prairie-Chicken recovery plans (1983, 1993, 2010) stress habitat protection and restoration in multiple geographically separated areas, but until 1990, the APCNWR was the only area where any habitat protection, restoration, or management had been accomplished. Habitat for the Attwater's Prairie-Chicken should (1) consist of at least 33 to 50 percent grassland; (2) be within 1.6 km of traditional/historical booming grounds; and (3) be burned before 1 February or mowed before 1 July to provide grass and residual cover long enough (25 cm), but not too long (≥ 1 m), for nesting and brood rearing.

Heavy spring floods in 2016 and Hurricane Harvey in 2017 were disastrous for the remaining wild populations of Attwater's Prairie-Chicken residing on APCNWR. Captive-breeding programs are underway at the Houston Zoo, Fossil Rim Wildlife Center near Glen Rose, Abilene Zoo, and Caldwell Zoo (Tyler, Texas). A flock reintroduced from the Houston Zoo's captive-breeding program resides on the grounds of NASA's Lyndon B. Johnson Space Center near Clear Lake.

Attwater's Prairie-Chicken, *Tympanuchus cupido attwateri*

FEDERAL STATUS: Endangered
TEXAS STATUS: Endangered

DISTINGUISHING FEATURES. Attwater's Prairie-Chicken is a smaller, darker subspecies of the Greater Prairie-Chicken. The upperparts are dark and tawny, with chestnut on the neck; the dark bars on the underparts are narrow and the light bars are buffy. The spots on the wing coverts are smaller than those of the Greater Prairie-Chicken, and the male's pinnae feathers are shorter. Attwater's Prairie-Chickens also have less feathering on the feet and tarsi (legs) than Greater Prairie-Chickens.

SPECIFIC DISTRIBUTION AND HABITAT. Attwater's Prairie-Chickens are endemic to the coastal prairies of Texas and Louisiana. They were formerly distributed along the Texas Gulf Coast and up to about 120 km (75 miles) inland from the Nueces River near Corpus Christi (San Patricio County) to Sabine Pass (Jefferson County), and east to the Bayou Teche region of Louisiana. Although they were historically found in as many as 48 counties in Texas, their population was estimated at 8,700 in 1937 and they were found in only 19 counties. By 1999 they numbered fewer than 200 and were found in only 2 counties. The subspecies is currently found only at the Attwater Prairie Chicken National Wildlife Refuge (APCNWR; Colorado County, Texas) and a private ranch in Goliad County, Texas, where it has been reintroduced (fig. 9.8). There are only about 90 birds left in the wild.

The Gulf Coastal Prairie, described as a "sea of grass" by early settlers, once encompassed more than 6 million acres. Preferred habitat for Attwater's Prairie-Chicken is large tracts of well-drained grasslands with some forbs and shrubs but few trees, especially near breeding areas. Open space is important, especially for leks, but distribution of woods and openings is also important for nesting. During the summer, young Attwater's Prairie-Chickens require ample shade and access to surface water.

The APCNWR was established in 1972, five years after the species was listed but more than 10 years before the first recovery plan was completed. Currently, restoration efforts are focused in three areas where relatively large Attwater's Prairie-Chicken populations occurred historically: (1) the Texas City Prairie Preserve in Galveston County, managed by The Nature Conservancy; (2) the Austin–Colorado County priority management zone, which includes the APCNWR; and (3) the Refugio–Goliad County priority management zone. The Texas City area is undergoing rapid urbanization, limiting its long-term potential. The Austin–Colorado County priority management zone contains 80 percent of the booming grounds that existed there from 1979 to 1992. The Refugio–Goliad County priority management zone (268,690 ha) contains the largest contiguous blocks of coastal prairie remaining in Texas and about 31 percent of the bird's historical booming range.

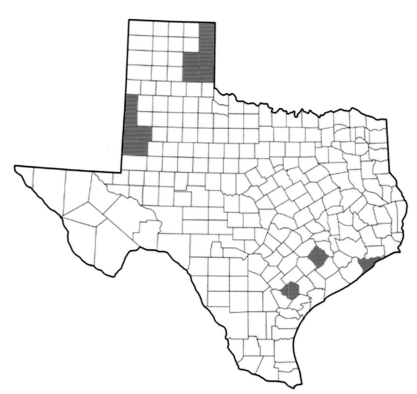

Figure 9.8. The remaining Attwater's Prairie-Chickens exist in three geographically separated counties in the Gulf Coastal Prairie (blue). Two geographically separated metapopulations of Lesser Prairie-Chickens occupy the High Plains in the Texas Panhandle (red). Map by Brian R. Chapman.

FEDERAL DOCUMENTATION:

Listing: 1967. Federal Register 32:4001

Critical Habitat: Not designated

Recovery Plan: US Fish and Wildlife Service. 2010. Attwater's Prairie-Chicken recovery plan, 2nd revision. Albuquerque, NM. 107 pp.

Lesser Prairie-Chicken,
Tympanuchus pallidicinctus

FEDERAL STATUS: Under Review (Threatened 2014, rescinded by court order 2016)

TEXAS STATUS: Not Listed

DISTINGUISHING FEATURES. Lesser Prairie-Chickens were first described as a subspecies of the Greater Prairie-Chicken. Their appearance differs little from that of the Greater Prairie-Chicken; most notably, during the breeding season their gular sac is predominantly red compared to the orange of

Greater Prairie-Chickens (fig. 9.9). In addition to habitat differences, Lesser Prairie-Chickens also spend more of the year on the booming grounds than Greater Prairie-Chickens, have a slightly different courtship display, and have higher-pitched booming notes that are less "ventriloquial."

SPECIFIC DISTRIBUTION AND HABITAT.
Historically, the range of the Lesser Prairie-Chicken included extreme eastern New Mexico and the Texas Panhandle south and east through the western Edwards Plateau and into the Permian Basin, western Oklahoma (including the Panhandle), much of southern Kansas, and the southeastern corner of Colorado. Today, the range is restricted to extreme southeastern Colorado and southwestern Kansas, the Oklahoma Panhandle, a few counties in northwestern Oklahoma and the adjacent Texas Panhandle, and an area of southeastern New Mexico and the southwestern Texas Panhandle (see fig. 9.8). In Texas, there are two metapopulations in two disjunct areas encompassing 11 counties: Lipscomb, Hemphill, Wheeler, Collingsworth, Donley, and Gray (the northeastern Panhandle, extending into Oklahoma); and Bailey, Cochran, Yoakum, Gaines, and Terry (the southwestern Panhandle, extending into New Mexico).

Lesser Prairie-Chicken habitat is described as "rolling, sandy, bunchgrass country." The birds appear to select large areas of suitable habitat that are based on vegetation structure rather than composition and include mid to tall grasses and shrubs. Preferred habitats are dwarf shrub–mixed grass prairies of bluestem, grama, and threeawn

grasses and Sand Dropseed interspersed with scattered clumps of Sand Sage or Shinnery Oak. In the western portion of their range, sandy soils that will support grass and shinnery rather than shortgrass prairie are required. Booming grounds are typically on slightly elevated areas with sparse vegetation. New leks may be established in areas where disturbances have reduced vegetation density or height; other disturbances, such as fire or tillage, may cause traditional leks to be abandoned. Lesser Prairie-Chicken hens often nest beneath dwarf shrubs or tall bunchgrasses. Taller trees and shrubs are used for shade during summer.

FEDERAL DOCUMENTATION:
Listing: 2014. Federal Register 79:19974–20071 (Threatened)
2016. Federal Register 81:47047–47048 (Rescinded)
2016. Federal Register 81:86315–86317 (Status Review Initiated)
Critical Habitat: Not designated
Recovery Plan: Van Pelt, W. E., S. Kyle, J. Pitman, D. Klute, et al. 2013. The Lesser Prairie-Chicken range-wide conservation plan. Western Association of Fish and Wildlife Agencies, Cheyenne, WY. [The USFWS "endorsed" this version of the Western Association of Fish and Wildlife Agencies' Lesser Prairie-Chicken range-wide conservation plan. Available at http://www.wafwa.org/Documents %20and%20Settings/37/Site%20Documents /Initiatives/Lesser%20Prairie%20Chicken /2013LPCRWPfinalfor4drule12092013.pdf.]

Mexican Spotted Owl,
Strix occidentalis lucida

JOHN KARGES

FEDERAL STATUS: Threatened
TEXAS STATUS: Threatened

The Mexican Spotted Owl is a rarely seen and uncommon bird of evergreen forests in steep canyons of mountainous "sky islands" in the southwestern United States and Mexico. In Texas, only a few pairs reside in higher mesic wooded canyons and forest patches in the Guadalupe Mountains, the tallest Texas mountain range, and fewer still, if any, remain in the Davis Mountains, the second tallest. The Mexican Spotted Owl is poorly studied and rarely monitored in Texas because both populations are outliers, and most of the species' range is in the southwestern United States.

DESCRIPTION. Spotted owls are medium-sized owls with dark eyes. They lack the "ear tufts" characteristic of some other owl species but are similar to others in having a rounded head. They are most similar in shape and coloration to the Barred Owl, which occurs throughout much of the eastern third of Texas. Overall, the Mexican Spotted Owl is dark brown dappled with white, and the breast and belly are adorned with the conspicuous oval white spots for which it is named. The facial disks are dark, with a distinctive lighter X pattern centered above the beak (fig. 9.10). The sexes are virtually indistinguishable by appearance. The species has as many as 13 different vocalizations serving different social needs and behaviors, including hoots, barks, and whistles. Most notable, and often heard, is a series of multiple hoots given by territorial males and females or by males returning to nests with food.

DISTRIBUTION. Spotted owls occur from the Pacific Northwest in British Columbia southward along the Pacific Seaboard mountain ranges into Baja California, the Sierra Nevada of California, and throughout mountain ranges of the southern Rockies in central Utah and Colorado southward through the isolated "sky island" mountain ranges of Arizona and New Mexico to the Sierra Madre Oriental, Sierra Madre Occidental, and the Sierra Madre del Sur of southern Mexico. There are three subspecies. The Northern Spotted Owl occurs in the upper Pacific Northwest. The California Spotted Owl occupies the California coast from the Bay Area into the higher mountain ranges of Baja California and into the Sierra Nevada. The Mexican Spotted Owl inhabits the southern Rocky Mountains southward to the Sierra Madre del Sur and the Transverse Volcanic Ridge in southern Mexico.

Figure 9.10. Mexican Spotted Owl. Photograph courtesy of the US Fish and Wildlife Service.

HABITAT. The species requires mature forests with dense canopies. The forests include evergreen trees in old-growth Pacific Northwest mesic forests, oak woodlands in California and Baja California mountains, and forests dominated by evergreen or coniferous species in steep canyons or against rocky cliffs and bluffs in the southern Rockies, the sky islands of the southwestern United States and northern Mexico, and the Sierra Madre of Mexico.

NATURAL HISTORY. Spotted owls are mostly sedentary, making only limited and short migrations between wintering and breeding sites, often between foothill lowlands and montane forests where they breed. They are territorial with nearby pairs or adults. The male selects the nest site, which may be a rocky ledge, a protected shelf under an overhang, the snag of a broken tree trunk, a hollow tree cavity well above the ground, Mistletoe clumps ("witch's brooms"), or an old raptor or squirrel nest (although there are no arboreal squirrels in either Texas mountain range). The female prepares the nest by scraping debris and adding some feathers for nest lining, but the nest is often crude and minimal. A pair may use a nest site for multiple nest attempts through the years, but they use it only once a year, and not every year (most pairs do not nest every year). Clutch size varies from one to four eggs. The eggs are incubated for a month and the juveniles fledge about a month after hatching. Juvenile mortality is high, but adult survival is also high. Youngsters remain with their parents for several weeks after fledging and disperse from their natal territory in early fall. The adults are solitary except with mates and young and form long-term monogamous pair bonds. A mated pair occupies its territorial home range for several years.

Spotted owls are deft and maneuverable flyers, well adapted for the dense forests where they occur. They forage in the forest and detect their prey of small to large rodents by sight and sound. They most often capture packrats and pocket gophers but take other vertebrate prey, including bats on the wing. Almost all hunting is done at night from dusk to dawn.

Most predation on adult and juvenile Mexican Spotted Owls in Texas is most likely by Great Horned Owls. The most common predator of eggs and hatchlings is the Common Raven.

REASONS FOR POPULATION DECLINE AND THREATS. In Texas, the major factor limiting the distribution and abundance of the owl is the small area of suitable habitat concentrated within a restricted geography. The only places Mexican Spotted Owls have ever been known to nest are in the Guadalupe and Davis Mountains sky islands, isolated ranges with sufficient elevation and rainfall to support the Madrean woodlands (evergreen-oak forests) the owls require. The Guadalupe Mountains National Park population is not surveyed every year but is reported to contain only a few pairs; no resident pairs are currently known in the Davis Mountains. There are only scattered reports of them being seen since the early 1990s, and their existence in the mountains has not been recently confirmed. Surveys in the Davis Mountains are restricted to lands available to researchers. Large areas containing appropriate nesting habitat are privately owned

and inaccessible. Thus, a full survey or monitoring effort for the entire mountain range is precluded.

Potential immigration or emigration (exchange and recruitment) between Mexican Spotted Owls in the Davis Mountains with the nearest neighboring population in the Guadalupe Mountains, a distance spanning approximately 70 miles, is unlikely but unknown. No suitable habitat exists for a "bridge" or wooded corridor between the ranges. There may be a greater possibility for population exchanges between owls in the Guadalupe Mountains and the next sky island to the north, the Sacramento Mountains of south-central New Mexico. The Mexican Spotted Owl is reported but unconfirmed in the Chisos Mountains in Big Bend National Park and is not known from the Maderas del Carmen, the forested sky island south of Big Bend National Park in the adjacent state of Coahuila, Mexico.

A concern for the conservation of the Mexican Spotted Owl in Texas is changes in the forest health and contiguity of forested expanses within the mountains, particularly in the Davis Mountains, where climate change and historical land use are intertwined in altering forest health and vitality with water stress and the potential loss of mature pine and oak in the mesic canyon woodlands. Combined conditions of forest overcrowding, stand replacement, wildfire, Pine Beetle outbreaks and infestations, and poor recruitment of pines following droughts, fire, and other landscape perturbations all contribute to habitat loss and threaten the species. Because the birds are not monitored or tracked in either mountain range, no one knows what becomes of dispersing offspring when they fledge.

RECOVERY AND CONSERVATION. Only the Guadalupe Mountains population was considered in the recovery plan (2012) as a portion of the Basin and Range East Ecological Management Unit in the subdivisions of the range within the United States, because it is close to the larger population in the Sacramento Mountains of New Mexico. Its federally threatened status is given attention in resource and fire management plans developed for Guadalupe Mountains National Park by the National Park Service and for the adjacent Lincoln

National Forest in New Mexico by the US Forest Service. No agency fire management plan includes consideration of the Davis Mountains birds, but The Nature Conservancy, which owns the Davis Mountains Preserve, includes stewardship and management of habitat in its plans where the owls have been sighted.

Texas locations were not included in the critical habitat designation for the species because the Texas owl population was too small to be considered essential to the recovery of the species. Critical habitat was designated only in Utah, Colorado, New Mexico, and Arizona. The small populations in south-central New Mexico, those closest to the subspecies' range in Texas, were likewise omitted.

FEDERAL DOCUMENTATION:

Listing: 1993. Federal Register 58:14248

Critical Habitat: 2004. Federal Register 69:53181

Recovery Plan: US Fish and Wildlife Service. 2012.
Final recovery plan for the Mexican Spotted Owl (*Strix occidentalis lucida*), first revision. Albuquerque, NM. 413 pp.

Figure 9.11. Northern Aplomado Falcon. Photograph by Robert Burton, US Fish and Wildlife Service.

Northern Aplomado Falcon,
Falco femoralis septentrionalis

TIMOTHY BRUSH

FEDERAL STATUS: Endangered
TEXAS STATUS: Threatened

The Northern Aplomado Falcon was extirpated from the northern portion of its extensive range by the 1950s. After several decades of absence from Texas, the species was reintroduced into its former breeding range, and two populations in South Texas are slowly expanding.

DESCRIPTION. The Northern Aplomado Falcon is a colorful, long-winged, long-tailed raptor. The crown, most of the head, and the upperparts are steel gray (the Spanish word *aplomado* means "lead-colored"). A white stripe extends back from the eye, and a dark gray "mustache" sinks downward through the white plumage of the neck. A wide, dark band separates a white or buffy upper chest from a rufous to cinnamon belly (fig. 9.11). The white-tipped tail is dark but barred with white or grayish bands and black stripes. In flight, the trailing edges of the wings are white, contrasting sharply with the dark wing linings. Adults are 14 to 18 inches long and have a wingspan of 31 to 40 inches.

DISTRIBUTION. The Coastal Prairie of South Texas and the Chihuahuan Desert grasslands from West Texas to southern Arizona historically represented the northern periphery of the Aplomado Falcon's breeding range. The species is widely distributed through Mexico, Central America, and South America to the southern tip of the continent. The falcon declined substantially in coastal Texas between 1890 and 1910 and was extremely rare across its former Texas range from 1910 to the 1950s. The last known natural US breeding records were in South Texas in 1941 and in New Mexico in 1952. A few individuals wandered into West Texas in the 1990s from a presumed remnant population in northern Mexico. After years of recovery efforts beginning in the 1980s that involved captive breeding and reintroductions, the species now inhabits the Texas Coastal Prairie from

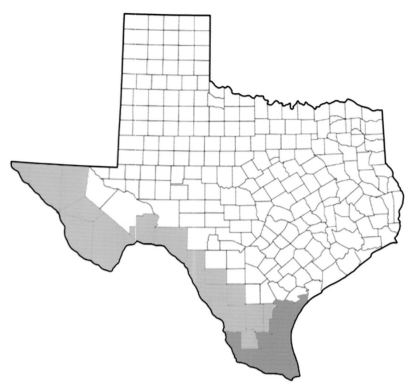

Figure 9.12. Green denotes the area currently occupied by descendants of reintroduced Northern Aplomado Falcons; yellow approximates their historical range. Map by Brian R. Chapman.

Brownsville to Rockport (fig. 9.12). Reintroductions to the desert grasslands in the Trans-Pecos region were unsuccessful, but individuals occasionally wander into West Texas from New Mexico.

HABITAT. In Texas, the Northern Aplomado Falcon inhabits open desert grasslands and coastal prairies studded with scattered cacti, yuccas, and Honey Mesquite. In the Rockport area, the falcon sometimes perches on telephone wires and Live Oak boughs.

NATURAL HISTORY. The Northern Aplomado Falcon feeds opportunistically on a variety of prey, including large insects, lizards, snakes, birds, bats, and small rodents. Although most searches for prey occur during the daylight hours, some begin hunting for prey well before daylight and may extend their quest after sunset. The falcon locates prey while soaring, flying rapidly through dense shrubs or trees, or perching on a high observation post. It pursues birds in flight and sometimes

chases rodents on the ground. Mated pairs and sometimes larger groups cooperate as a team to flush and capture prey.

Courtship begins in the spring, often coinciding with a search for a nest. The pair does not construct a new nest but instead seeks abandoned nest platforms constructed by large birds. Most nests are flimsy, flat or conical structures placed on yuccas, power-line cross arms, or cliffs. After the clutch of three (rarely two or four) white or buffy eggs with scattered rusty spots is laid, the pair cooperates in incubation for 31–32 days; the female spends more time in incubation than the male. When the chicks hatch, they are covered in white down and are helpless. The male brings food to the nest, and the female usually feeds the young during the four to five weeks before fledging. When the birds leave the nest, their flight feathers are not fully developed, so the fledglings remain nearby for several weeks and are fed by their parents. Captive Aplomado Falcons have lived for up to 24 years, but wild birds do not often live beyond 5–6 years.

REASONS FOR POPULATION DECLINE AND THREATS. A significant decline in the Texas populations of the Northern Aplomado Falcon occurred between 1890 and 1910. Declines during this period pre-dated widespread conversions of prairies to agriculture, although brush encroachment into prairies, caused by overgrazing, may have been a factor. Continued declines of remaining birds in Texas in the early twentieth century may have been related to agricultural conversion and continued brush encroachment, as well as heavy pesticide use. After World War II, liberal applications of DDT, dieldrin, and other organochlorine pesticides drastically reduced nesting success. These chemicals, which persisted in the food chain, were found in heavy concentrations in Aplomado Falcons in eastern Mexico. By the mid-1950s, the Northern Aplomado Falcon was extirpated from the state.

RECOVERY AND CONSERVATION. The Peregrine Fund released 927 young Aplomado Falcons into coastal prairies of the Lower Rio Grande Valley and the Coastal Bend region from 1978 to 2013. To ensure nesting success, some landowners provided nesting platforms modified to reduce predation on nestlings by Great Horned Owls. The Peregrine Fund also released 637 captive-reared falcons at 11 sites in West Texas from 2002 to 2011, and a small nesting population developed. However, an extended drought in the region eliminated this population, and the reintroduction program there was discontinued. The current breeding population of Northern Aplomado Falcons along the southern and central Texas coast represents descendants of captive-reared birds. The continued recovery of the species in South Texas appears promising, with 30–33 pairs in 2008–2013, including 15–18 near Brownsville and 15 on islands in the Coastal Bend region. Estimates in 2017 were similar.

FEDERAL DOCUMENTATION:

Listing: 1986. Federal Register 51:6686–6690
Critical Habitat: Not designated
Recovery Plan: US Fish and Wildlife Service. 1990. Northern Aplomado Falcon recovery plan. Albuquerque, NM. 56 pp.

Sprague's Pipit, *Anthus spragueii*

CLIFFORD E. SHACKELFORD

FEDERAL STATUS: Candidate
TEXAS STATUS: Not Listed

Populations of grassland birds have declined more rapidly than those of any other avian guild. Many of these species, including Sprague's Pipit, are short-distance migrants. After breeding in the northern Great Plains, these birds winter in the grasslands of South Texas, the southwestern United States, and northern Mexico. Conversion of native prairies for agriculture and urban development across both the breeding and wintering grounds has resulted in a 75 percent population decline since the mid-1960s.

DESCRIPTION. Sprague's Pipit is a small, drab, earth-toned passerine that is about 4–6 inches long, with brown wings and tail and a buffy breast streaked with black. Similar in appearance, both males and females have two faint wing bars and white outer rectrices (tail feathers). A pale white

Figure 9.13. Sprague's Pipit. Photograph by Dominic Sherony, Creative Commons.

eye ring makes the eyes appear large against a buffy background, which extends to the crown and nape (fig. 9.13). The upper mandible of the slender, straight bill is blackish, while the lower mandible is pale with a blackish tip. The Sprague's Pipit's streaked back and pink or yellowish legs distinguish it from the American Pipit, which has an unstreaked back and dark legs.

This bird usually skulks low in the grass, staying hidden from view, but sometimes stretches its head up to peek at potential approaching danger. The species frequents edges of gravel roads bordering or bisecting open pastures. Its most distinguishing feature is behavioral—how it flees when flushed. After takeoff, it "stair-steps" into the air, chirping along the way before circling high above. It then dives abruptly to the ground, often landing close to where it initially flushed. No other species in Texas flushes in this manner. Unlike the American Pipit, which it resembles, Sprague's Pipit does not bob its tail, occur in small flocks, or visit exposed grassy or muddy openings.

DISTRIBUTION. The breeding range of Sprague's Pipit includes the grasslands and aspen parkland regions of the Canadian prairie provinces (Alberta, Manitoba, and Saskatchewan). The northern Great Plains of the United States, particularly the central mixed-grass prairies of Montana, North Dakota, and South Dakota, support the greatest densities, but the breeding range extends to northwestern Minnesota. Most Sprague's Pipits winter from Central and South Texas south to central Mexico. Fewer wintering birds occupy the peripheral portions of the range, which extends westward to southeastern Arizona and southern New Mexico and eastward to southern Louisiana.

HABITAT. In both breeding and wintering ranges, Sprague's Pipits display strong affinities to intermediate-height grasslands that have never been plowed. Preferred habitat conditions include well-drained, open prairies where grass heights are between 4 and 12 inches and few shrubs or trees interrupt the expanse. Planted nonnative grasslands with similar characteristics also are used. During migration, Sprague's Pipits occasionally visit roadside or trailside grassy areas, but these habitats, grass strips bordering agricultural fields, cultivated areas, and fallow, mowed, or recently burned meadows with short vegetation are generally avoided at other times.

Although winter habitats are inadequately described, densities of wintering birds tend to be greater in thick grassy habitats. Small areas of bare ground embedded within expansive grasslands appear to be important habitat components. Such openings may be created by certain edaphic conditions, periodic grazing, or other causes.

NATURAL HISTORY. After arriving on the breeding grounds in April through mid-May, males establish territories in dense grassy habitats. Their high aerial flight displays may last for up to three hours each day. Territories are established in native grassland patches exceeding 358 acres; the pipits generally avoid smaller patches and do not occupy those smaller than 72 acres. Most territories are centered in elevated areas with uninterrupted expanses of short grass containing few sedges or forbs. Females weave dome-shaped nests from grasses and situate them amid dense cover at the end of a long, covered, sharply curved runway in dense grass. Clutches contain four to five eggs, and after hatching, both parents feed the young and continue parental care well past fledging.

Most Sprague's Pipits forage alone, gleaning food from the ground or low vegetation as they walk through the grass. During the breeding season, arthropods make up the bulk of the diet, but pipits also eat some vegetable matter. Seeds become a more important part of the diet during the winter.

Migration from the northern portions of the breeding range to the wintering grounds begins in August. Apparently, some migrants make brief stops as they travel through eastern New Mexico, Oklahoma, or North Texas. The greatest densities of wintering Sprague's Pipits are concentrated in South Texas. While on the wintering grounds, the birds do not vocalize much except when flushed. The species produces a distinctive high-pitched, chirpy flight call that is easily overlooked.

REASONS FOR POPULATION DECLINE AND THREATS. Christmas Bird Count data for the 40-year period from 1966–1967 through 2005–2006 show that the overwintering Texas population declined 2.54 percent annually and 73 percent overall for the time period. Loss, alteration, and fragmentation of native grasslands in the breeding and wintering ranges have reduced Sprague's Pipit recruitment and survivorship. Only about 2.1 percent of the grassland habitat within the breeding range in the United States and approximately 6 percent of the breeding range in Canada remains as suitable habitat. Most of the historical range has been converted to agriculture or other uses, and the rate of habitat conversion and fragmentation is accelerating. Transformation of native prairies to nonnative "tame" pastures, woody encroachment from lack of prescribed fire, and human development also convert preferred grasslands into completely unsuitable habitats. Other documented contributors to population declines include nest predation and parasitism, drought, and invasion by nonnative plants. Potential threats include the expansion of oil and gas production and wind energy development in prairie habitats.

RECOVERY AND CONSERVATION. In 2009 and again in 2015 the US Fish and Wildlife Service (USFWS) determined that the listing of Sprague's

Pipit as threatened or endangered was warranted. The agency chose not to proceed with the listing process because the action was precluded by higher-priority listings. Consequently, a recovery plan was not developed, but the USFWS published a conservation plan with a similar format. Sprague's Pipit is listed as a priority species in the Texas Parks and Wildlife Department's Texas Conservation Action Plan.

Protection, management, and restoration of large grassland tracts are essential for conserving Sprague's Pipit on both its breeding and wintering grounds. On the Coastal Prairie of Texas, methods such as herbicide applications and prescribed burning to control invading mesquite and other shrubby vegetation are crucial for maintaining the mixed-grass habitats preferred by wintering grassland birds. Much additional information about the winter habitat requirements of Sprague's Pipit is needed to focus conservation efforts in Texas on preservation of the few remaining large grassland tracts.

FEDERAL DOCUMENTATION:

Listing: 2010. Federal Register 75:56028–56050
Critical Habitat: Not designated
Conservation Plan: Jones, S. L. 2010. Sprague's Pipit (*Anthus spragueii*) conservation plan. US Fish and Wildlife Service, Washington, DC. 56 pp.

Southwestern Willow Flycatcher,
Empidonax traillii extimus

BRIAN R. CHAPMAN

FEDERAL STATUS: Endangered
TEXAS STATUS: Endangered

The Southwestern Willow Flycatcher was never numerous in Texas, and its current status in the state remains unknown. If this endangered subspecies nests in West Texas, individuals or small breeding groups likely occupy widely scattered patches of habitat along rivers. Canyons where springs or deep depressions provide semipermanent aquatic environments with dense vegetative margins may also offer suitable nesting habitats.

Figure 9.14. Southwestern Willow Flycatcher. Photograph courtesy of the US Fish and Wildlife Service.

DESCRIPTION. Flycatchers of the genus *Empidonax* are so similar in appearance that experts sometimes err when attempting to identify (by sight alone) any of the seven species that occur in Texas. Willow flycatchers are small, sparrow-sized birds with dull grayish-olive upperparts, a pale yellowish belly, a whitish throat, two dull white wing bars, and a faint and easily overlooked white eye ring (fig. 9.14). The bottom mandible of the bill is yellow, but it sometimes appears dark when viewed from a distance. The species can be distinguished by voice—the distinctive song is a wheezy *fitz-bew* or a rising *breeet*—and behavior—a habit of flicking the tail upward when perched. Four subspecies of willow flycatcher are currently recognized, but recent evidence suggests that the Southwestern Willow Flycatcher is not a valid subspecies.

DISTRIBUTION. The breeding range of the Southwestern Willow Flycatcher extends from southern California eastward through the southern portions of Nevada, Utah, Arizona, New Mexico, and into western Texas. The subspecies may occur in any of nine Texas counties west of the Pecos River (fig. 9.15). Adult Southwestern Willow Flycatchers occupy the breeding range from early May until late August, but juveniles may linger until late September.

Figure 9.15. Potential distribution of the Southwestern Willow Flycatcher in Texas. Map by Brian R. Chapman.

loose material dangling from the bottom. Nest height varies from 1.6 to 26 feet.

Females lay clutches of three to four eggs and begin incubation, which requires 12–15 days, after depositing the last egg. Nestling growth and development is rapid, and the young birds usually leave the nest (fledge) within 15 days of hatching. Fledglings remain near the nest, where the parents feed them for about 15 more days. Should a nest be abandoned because of predation, parasitism, or disturbance, willow flycatchers usually make a second nesting attempt, sometimes with a new mate. Second nests are constructed more rapidly and the clutches typically have fewer eggs.

Sometimes described as a "sit-and-wait" predator, these flycatchers usually sally forth from a perch to grab passing insects. Willow flycatchers also feed by gleaning, which involves hovering to pluck insects from leaves or other plant parts, and by capturing insects crawling on the ground. Prey includes flying ants, flies, wasps, bees, and beetles. Berries, seeds, and small fruits also are dietary components, but they are not a significant food source.

HABITAT. Habitat preferences in Texas have not been reported. In other portions of its range, the flycatcher typically constructs nests in riparian habitats having a dense canopy and an equally dense, shrubby understory.

NATURAL HISTORY. The Southwestern Willow Flycatcher is migratory. Migration to the wintering grounds may begin as early as mid-July, and migratory movements extend through September. The winter range includes southern Mexico, Central America, and western Venezuela.

Shortly after arriving at a suitable breeding habitat in early May, males establish and defend a territory by singing. Song perches, located throughout a territory, vary from high, visible locations to branches concealed within dense vegetation. Some territories are in isolated habitat patches large enough for only one or a few pairs. Females construct open cup nests in the forks of branches by weaving together grass, leaves, animal hair, and other fibrous materials. Nests often have

REASONS FOR POPULATION DECLINE AND THREATS. Extensive loss, fragmentation, and modification of riparian and other wetland habitats have reduced breeding habitat throughout its range. Habitat alterations, especially the degradation of cottonwood-willow riparian associations historically used as nesting habitats, may result from agricultural development, drawdown of local water tables, water diversions and impoundment, channelization, fires, overgrazing by cattle, and replacement of native vegetation by nonnative species such as Saltcedars.

Influxes of predators, such as snakes, mammals, and predatory birds, and increases in nest parasitism by Brown-headed Cowbirds often accompany habitat alterations associated with agricultural development. Brood parasitism sometimes plays a major role in population declines, and when parasitizing a flycatcher nest, cowbirds often function as predators when they remove flycatcher eggs or young.

RECOVERY AND CONSERVATION. Individual flycatchers exhibit greater fidelity to a local area than to a specific nest location and often move annually within a drainage basin or to nearby drainage basins. Riparian zones designated as critical habitat provide sufficient wetland vegetation for breeding, nonbreeding, territorial, dispersal, and migratory activities. The recovery plan does not designate any critical habitat in Texas because suitable habitats are widely dispersed and the Southwestern Willow Flycatcher may no longer exist in the state. Habitat that appears suitable for breeding sites exists within the boundaries of Big Bend National Park, Big Bend Ranch State Park, and Black Gap Wildlife Management Area. These federal, state, and private entities protect suitable habitats in Texas.

FEDERAL DOCUMENTATION:

Listing: 1995. Federal Register 60:10694–10715

Critical Habitat: 2013. Federal Register 78:343–534

Recovery Plan: US Fish and Wildlife Service. 2002. Final recovery plan for the Southwestern Willow Flycatcher. Albuquerque, NM. 229 pp.

Western Yellow-billed Cuckoo,
Coccyzus americanus occidentalis

HEATHER MATHESON

FEDERAL STATUS: Threatened
TEXAS STATUS: Threatened

There is an ongoing debate among ornithologists concerning the biological distinctiveness of the Yellow-billed Cuckoos in western states and those in the eastern United States. The conservation status of the western population, however, is not a subject of debate. Populations of the Yellow-billed Cuckoo have declined throughout their range, but the most precipitous declines have occurred in the western half of North America. Without referring to a western subspecies, as some ornithologists propose, the US Fish and Wildlife Service listed the western population segment of the Yellow-billed Cuckoo as threatened.

Figure 9.16. Western Yellow-billed Cuckoo. Photograph by Stephen Ramirez, Creative Commons.

DESCRIPTION. The Yellow-billed Cuckoo is a slender, 12-inch-long bird with a long tail that has distinctively bold white spots on the underside. The plumage is brownish gray above and white below. In flight, the wing feathers appear reddish brown. As in all cuckoos, two toes at the end of the bluish-gray legs point forward and two point backward. Adults possess a narrow yellow eye ring, and the lower part of the down-curved bill is often yellow (fig. 9.16). Their call, often produced in response to loud noises such as thunder, is a distinctive croaking chuckle that tapers off at the end.

DISTRIBUTION. Historically, the breeding range of the Yellow-billed Cuckoo extended across North America from southern Canada to northern Mexico. More recently, the western population was extirpated from British Columbia, Washington, and Oregon. The remaining population segments in southwestern states are highly fragmented and small. Isolated populations may remain in Brewster, Culberson, El Paso, Hudspeth, Jeff Davis, or Presidio Counties in West Texas, but their current status is unknown.

Yellow-billed Cuckoos migrate to South America to spend the winter months. The western birds likely migrate down the western slope of Mexico and through Central America. They may take a different route during spring migration, skipping Central America and instead migrating across the Caribbean. Furthermore, evidence from geolocators indicates that Western Yellow-billed Cuckoos traverse south-central and North Texas during migration.

HABITAT. In the western states, Yellow-billed Cuckoos most often inhabit large blocks of riparian woodland where dense stands of willows, cottonwoods, or mesquite exist. These stands of riparian vegetation might occur along natural waterways and streams, as well as around developed wells, reservoirs, or other human-made riparian areas. Potential habitat usually includes more than 7 acres of continuous riparian vegetation cover. Nesting habitat in Texas is poorly known, but in the western states, willows seem to offer preferred nest sites while nearby cottonwoods provide foraging habitat. Cuckoos seem to prefer rough topography and areas with dense, native, heterogeneous forest around a core area of about 170 acres. Cuckoos are less likely to occur in vegetation dominated by nonnative Saltcedar.

NATURAL HISTORY. Both members of a nesting pair cooperate in constructing a nest of loosely intertwined sticks, incubating the eggs, and brooding the nestlings. Eggs are laid asynchronously, and the clutch size is typically two eggs but can range up to five. Eggs are laid within 24 hours of nest completion, but in some cases it can take up to five days. Incubation begins when the first egg is laid, so the first nestling emerges days before the second. Incubation typically lasts for eight to ten days but is variable. The male often sits on the nest during the night, while the female attends to the nest during the daytime; however, this pattern is not consistent among all nesting pairs. Nests are not left unattended for longer than 10 minutes during incubation or the early nestling period. When they hatch, the young are naked but alert. Both parents care for nestlings, but the effort by females decreases as the young age. Males often finish caring for nestlings and fledglings while females move away to nest again. Young develop quickly in the nest, typically leaving it around eight days after hatching. Although they cannot fly at this stage, they climb about in the canopy near the nest. Additional males or females sometimes help the pair tend to the nest.

Yellow-billed Cuckoos occasionally lay eggs in the nests of their own species and other species. The rapid growth of their nestlings gives them a developmental advantage over the nestlings of the host. However, the Yellow-billed Cuckoo is not an obligate brood parasite.

Yellow-billed Cuckoos feed primarily on insects, especially caterpillars, but also consume spiders, lizards, frogs, berries, and seeds. The birds feast during annual eruptions of hairy caterpillars, katydids, cicadas, and crickets.

REASONS FOR POPULATION DECLINE AND THREATS. Conversion of riparian habitats to farmland and housing developments is the primary cause for population declines in the western states. Modification of natural hydrological processes through stream channelization, water diversion, construction of roads or rail systems along rivers, and livestock grazing also contribute to habitat loss. Damming of rivers has altered riparian system dynamics, resulting in a lack of water for vegetation growth or extreme flooding of waterways during water releases from dams. Invasion of Saltcedar has contributed to the loss of native riparian vegetation, as have extreme and persistent droughts. During their long-distance nocturnal migrations to and from South America, the birds are vulnerable to collisions with a variety of tall structures. Little is known about threats during the winter.

RECOVERY AND CONSERVATION. Current efforts to recover the Western Yellow-billed Cuckoo are focused on management and restoration of native riparian habitat. These efforts include replanting native vegetation, protecting existing habitat, promoting propagation of existing vegetation, eliminating pesticide use in adjacent orchards, removing Saltcedar, and restoring natural hydrological processes. Patches of native vegetation should be heterogeneous and of the size (~170 acres) identified as required by cuckoos.

FEDERAL DOCUMENTATION:
Listing: 2014. Federal Register 79:59991–60038
Critical Habitat: 2014. Federal Register 79:71373–71375
Recovery Plan: Not developed

Black-capped Vireo, *Vireo atricapilla*

DAVID A. CIMPRICH

FEDERAL STATUS: Delisted due to recovery (2018)
TEXAS STATUS: Endangered

The Black-capped Vireo is a small, active songbird named for the black plumage of the male's head. This songbird is more often heard than seen as it sings loudly from within the thick vegetation that dominates its habitat. In Texas, it ranges primarily across the Edwards Plateau and Cross Timbers region and into the southern Trans-Pecos. Major breeding populations in the state occur at the Fort Hood Military Installation, in the western Hill Country at sites such as the Kerr Wildlife Management Area, and in the Devils River watershed.

DESCRIPTION. Black-capped Vireos are approximately 5 inches long and have white underparts, a green back, and pale yellow wing bars. The primaries (longest flight feathers of the wings), secondaries, and tail feathers are black with narrow green edges. Male head color is typically black, the throat is white, and white extends from the bill to just behind the eye (fig. 9.17). The sexes are similar except that the areas that are black on the heads of males are typically gray on females. Intermediates of both sexes occur that have gray crowns or napes and black foreheads, crowns, or cheeks. The breast color of females is pale buff whereas that of males is white.

Figure 9.17. Black-capped Vireo. Photograph by Greg Eckrich, US Fish and Wildlife Service.

The bill is gray and black, the legs are gray, and the iris is red, orange red, or red brown. Plumage color does not change seasonally.

DISTRIBUTION. The breeding range extends from Oklahoma through Texas to the Mexican states of Coahuila, Nuevo León, and Tamaulipas. Historically, the species also nested in south-central Kansas. Black-capped Vireos migrate to wintering areas along the Pacific slope of Mexico from southern Sonora to Oaxaca. Abundance appears greatest in the central portion of this winter range from southern Sinaloa to Colima. Migrating Black-capped Vireos have occurred at such widespread locations as Arizona, Ontario, and British Columbia.

HABITAT. Black-capped Vireos use a variety of habitats for nesting in Texas. A common feature of these habitats is low-stature deciduous shrubs with foliage 1.5–6.6 feet above the ground. Habitat usually consists of shrub patches interspersed with open areas of grass, forbs, or exposed rock. Here, strong light penetration supports the growth of foliage down to ground level around the edges of each patch. The species also nests in habitats where scattered trees grow above the shrubs and may even occur in woodlands with a shrub understory and few breaks in the tree canopy. In the northern and eastern portions of the species' range, the vireos usually occupy early successional habitats. In contrast, much of the vireo habitat in West Texas is climax vegetation, in which the low stature of woody species results from climatic or edaphic conditions. After becoming independent of their parents, young often move to areas of shrubs or forest bordering streams, rivers, and lakes.

NATURAL HISTORY. Males begin arriving on their Texas breeding grounds in mid to late March, followed by females a week or two later. Nesting begins in late March and April and ends by late July. Adults begin a complete replacement of their feathers shortly after their last nest of the season is no longer active. This molt is complete by the end of August. After molting, these birds deposit subcutaneous fat prior to southward migration. Because migratory departure occurs from late

August to early October, Black-capped Vireos spend approximately equal time on their breeding and wintering grounds.

During the breeding season, male Black-capped Vireos defend territories that average about 2.5 acres, but these may be as much as four times larger where populations have not saturated the available habitat. A male usually defends an area throughout the breeding season, only rarely relocating to a new territory midseason. Females may remain paired with the same male throughout the breeding season or switch to a new mate after the failure or success of a nesting attempt. In this way, the species can be either socially monogamous or sequentially polygamous. Males sing throughout the breeding season, with a marked decline as molt begins. When singing, males typically alternate between two or three songs selected from their large individual repertoires. Every several minutes, they replace one of the songs with a new selection. Territorial males sing throughout the day but have a particularly intense period of dawn singing just before sunrise that lasts 20–40 minutes.

Black-capped Vireos typically build their nests 0.5–5 feet above the ground, although nests as high as 13 feet have been recorded. They position their nests where a branch forks to form a Y or F shape. Supporting branches attach to the top of nests and are flexible, which allows an incubating female to feel the approach of a nocturnal nest predator in time to flee.

In response to repeated nest failures, Black-capped Vireos may continue to renest up to seven times within a single breeding season. In Texas, common nest predators include snakes, fire ants, birds, and mammals. Fewer than 20 percent of pairs attempt to raise a second brood after succeeding in fledging a first. Although clutch size ranges from one to five, most consist of three or four eggs, with three-egg clutches being more common late in the breeding season. Incubation lasts 13–14 days. Both males and females share incubation duty, but females usually spend longer periods on the nest than males, and only females incubate through the night. The nestling period lasts 10–12 days. Although both sexes bring food to the nest, males bring more than females. After the young leave the nest, they remain in the natal territory for five weeks.

REASONS FOR POPULATION DECLINE AND THREATS. Nest parasitism by Brown-headed Cowbirds decreases the reproductive success of the Black-capped Vireo. Cowbirds lay eggs in the nests of a variety of host species in order to have the host raise their young. Black-capped Vireos abandon about half of parasitized nests. When not abandoned, parasitized nests rarely produce vireo fledglings because eggs fail to hatch or nestlings die within a few days. During the breeding season, cowbirds shuttle between feeding areas where they forage in the company of cattle, and breeding areas where they parasitize nests. The two areas may be separated by as much as 4.5 miles. Cattle in proximity to breeding Black-capped Vireos can increase parasitism, and rates sometimes exceed 90 percent.

Habitat loss also threatens the Black-capped Vireo. Changes in land use can directly eliminate breeding habitat; intense browsing by deer, exotic wildlife, or goats can remove enough foliage in the height zone where vireos build nests to also render the habitat unsuitable. Suppression of wildfires removes the most important source of disturbance that creates or regenerates habitat for the species. Without periodic disturbance, the continued growth of shrubs and trees reduces low-growing foliage because of shading.

RECOVERY AND CONSERVATION. The recovery plan for the Black-capped Vireo was completed in 1991, when prospects for complete recovery were uncertain. Consequently, the plan established criteria for the interim objective of downlisting the species to threatened status. These criteria were that (1) existing populations be protected and maintained; (2) at least one viable population (> 500 pairs) exist in Mexico, Oklahoma, and four of six regions in Texas (later revised to four regions); (3) sufficient habitat be available on the wintering grounds to support the viable populations covered in criteria 1 and 2; and (4) all of these criteria be maintained for at least five consecutive years and the data indicate that they will continue to be maintained.

Listing of the Black-capped Vireo fostered extensive conservation efforts without which the species would likely be in greater jeopardy. Since the 1991 recovery plan was published, several Black-capped Vireo populations have increased considerably. At the Fort Hood Military Installation in Central Texas, researchers found 190 male vireos in 1992, but this number expanded more than 40-fold to an estimated 7,929 males in 2016. Similarly, the population in the Wichita Mountains National Wildlife Refuge and the adjacent Fort Sill Military Installation in Oklahoma increased from fewer than 100 males in 1987 to over 7,000 in 2016. These dramatic increases appear to be the result primarily of effective cowbird removal and secondarily of habitat management. For example, parasitism rates at Fort Hood decreased from 91 percent in 1987 to consistently less than 10 percent after 1996, and these low rates corresponded to increased reproductive success.

In addition to increases in existing populations, wild populations in some areas have been found to be larger than they were known to be when the species was listed. Such positive outcomes prompted the USFWS to recommend in 2007 that the species be downlisted to threatened. Following a 2016 status assessment, the USFWS further proposed that the species be delisted; the delisting decision was published in 2018.

FEDERAL DOCUMENTATION:

Listing: 1987. Federal Register 52:37420–37423
Delisting Announcement: 2018. Federal Register
    83:16228–16242

migratory bird with a breeding range restricted to the state. After spending the winter in pine-oak forests in southern Mexico and Central America, they return each year to breed in the juniper-oak woodlands of Central Texas. At the time of their listing as an endangered species, biologists estimated the entire population to be between 4,800 and 16,000 breeding pairs. Recent estimates suggest a larger, more robust population, but habitat losses on the breeding and winter ranges continue to impact this species. Golden-cheeked Warblers occupy juniper-oak woodlands from March through July at more than 20 public areas, including state, city, and county parks from San Antonio to Austin and the Balcones Canyonlands National Wildlife Refuge.

DESCRIPTION. Sightings of a Golden-cheeked Warbler bring delight to the human eye, but it may take patience to find the tiny bird. It is only about 5 inches long and weighs one-third of an ounce. This bird's beauty is striking. The entire side of the face is golden yellow, divided into an upper and lower portion by a black eye line (fig. 9.18). Adults have a black crown and throat, a black upper body, black wings with two white wing bars, a black tail with white outer rectrices, and a white belly. Females typically have an olive-black crown, may have little or no black on the throat, and may be grayer overall.

Male Golden-cheeked Warblers commonly use two different songs, designated the A and B songs. Both are loud (55 dB at 20 feet), with peak intensities at 5–6 kHz. The A song, which is used

Golden-cheeked Warbler,
Setophaga chrysoparia

JAMES M. MUELLER

FEDERAL STATUS: Endangered
TEXAS STATUS: Endangered

Loud songs sung from high perches in the Texas Hill Country announce the spring arrival of male Golden-cheeked Warblers laying claim to their territories. These small black-and-white birds with colorful yellow cheeks are native Texans—the only

Figure 9.18. Golden-cheeked Warbler. Photograph by Steve Maslowski, US Fish and Wildlife Service.

more frequently early in the breeding season, is shorter (about 1.5 seconds) and less complex than the B song, which is slightly longer (about 2 seconds) and more variable. Biologists use detection of these songs almost exclusively when surveying for the species during the breeding season.

DISTRIBUTION. The breeding range of Golden-cheeked Warblers extends from Palo Pinto County (about 60 miles west of Fort Worth) southwestward to Edwards County (about 125 miles west of San Antonio) and includes 36 counties in the Edwards Plateau and Cross Timbers regions of Texas. The heart of the winter range extends from Chiapas, Mexico, south through Guatemala, Honduras, El Salvador, and Nicaragua. The species occasionally ranges farther south to Costa Rica and northern Panama. Migration between the breeding and winter ranges traverses a 1,500-mile terrestrial route that follows the Sierra Madre Oriental through Mexico.

HABITAT. Golden-cheeked Warblers breed only in the juniper-oak forests of Central Texas. Some males establish territories in wooded areas as small as 7 acres, but contiguous habitat patches of at least 50 acres are necessary to achieve reproductive success rates that can sustain a population. Larger habitat patches tend to engender reproductive success. Closed-canopy forests where 70–80 percent of the trees are Ashe Juniper, Texas Red Oak, and Plateau Live Oak support the highest densities of nests. Reproduction is also successful in habitats that are more open, but nest success is lower in areas near forest edges. The Golden-cheeked Warbler, an insectivore, benefits from having a diversity of tree species within its territory. The populations of arthropods inhabiting various trees peak at different times and provide a smorgasbord of nutrition throughout the breeding season.

Golden-cheeked Warblers winter at elevations of 3,000–8,000 feet in montane pine-oak forests of southern Mexico and Central America. They forage for arthropods, primarily inside the outer foliage of oak trees in the forest midstory. Several oak species, especially the *encino*-type oaks with shiny,

narrow, elliptical or oblong leaves are preferred. During migration, the warblers visit many forested habitats within the Sierra Madre Oriental.

NATURAL HISTORY. The annual cycle of the Golden-cheeked Warbler begins each spring (late February through March) when males arrive on the breeding grounds to establish territories. Females arrive a few days after the males, select a nest site, and construct the nest. During this period, the territorial owner courts the female at the nest site by bringing her strips of Ashe Juniper bark removed from the main trunks of trees that are at least 20 years old (fig. 9.19). After weaving the bark into the nest and securing it with cobwebs and cocoons, the female allows the male to mate with her on the nest platform or nearby.

Nest construction requires about four days. The last step involves lining the nest bowl with grasses or other fine vegetation. During this period and until the chicks hatch, the male defends his territory by singing from perches, chasing intruding males, and joining the female when she leaves the nest to feed. Located 15 to 20 feet above the ground, the nests are difficult to find. They are small (roughly 3 inches wide and 2 inches deep), partially concealed in the forks of branches, and naturally camouflaged by the construction material.

The female begins laying eggs within three to four days of nest completion. Females lay one per day until the clutch of three or four eggs (rarely five) is complete. Incubation begins one day before the last egg is laid. Each egg weighs about 0.05 ounces, and chicks weigh nearly the same when they hatch, emerging with no feathers, sparse natal down, and partially opened eyes. Both adults feed the ravenous chicks, and over the course of nine days, the chicks increase their weight sixfold until they weigh nearly the same as an adult. Chicks fledge after 10–12 days. The brood often splits into two groups, and each parent feeds and protects the chicks as they wander around. About eight days after leaving the nest, the fledglings begin foraging on their own. The parents continue to provide food until they are completely independent, about four weeks after fledging.

By late summer, the adults and offspring begin

Figure 9.19. Strips of bark from old Ashe Juniper trees are used for nest construction by the Golden-cheeked Warbler. Photograph by Brian R. Chapman.

about 50 percent for adults; 1 in 1,000 birds lives 10 years. The oldest-known Golden-cheeked Warbler in the wild was 11 years old.

REASONS FOR POPULATION DECLINE AND THREATS. The primary threats to the Golden-cheeked Warbler are habitat loss and fragmentation resulting from urban encroachment and widespread clearing of juniper as a range management practice. In addition, habitat fragmentation has resulted in nest parasitism by the Brown-headed Cowbird.

Habitat loss is a continuing problem on both the breeding and winter range. An estimated 29 percent reduction of potential breeding habitat occurred in Texas between 2001 and 2011. This loss was especially rapid along the Balcones Escarpment from San Antonio to Austin and west of San Antonio, an area containing some of the largest expanses of potential habitat patches in the breeding range. Much of this loss is permanent because of urbanization. Nest parasitism by Brown-headed Cowbirds remains a potential threat, but preserving large blocks of breeding habitat can reduce rates of nest parasitism. Unfortunately, loss of habitat within the winter range continues at a rate of about 2 percent annually.

RECOVERY AND CONSERVATION. Breeding habitat has been protected using various measures. The Balcones Canyonlands National Wildlife Refuge, established in 1992, shields 25,745 acres of breeding habitat. The US Fish and Wildlife Service has approved several regional habitat conservation plans and conservation banks to protect large tracts of Golden-cheeked Warbler breeding habitat for perpetuity as mitigation for developments. The Balcones Canyonlands Conservation Plan, approved in 1996 for the city of Austin and Travis County, shields 31,780 acres of habitat for eight endangered species. Four other regional habitat conservation plans safeguard habitat in the rapidly developing areas along the Balcones Escarpment. Fortunately, the conservation of Golden-cheeked Warbler breeding habitat has found greater acceptance among private landowners who are interested in conserving the aesthetic value of

migrating south. This portion of the species' life cycle remains largely unknown. Although more than 6,000 Golden-cheeked Warblers have been banded in Texas since 1990, none of these has ever been resighted on the migratory path or in the winter range.

In winter, Golden-cheeked Warblers forage in mixed-species flocks in the montane pine-oak forests of southern Mexico and Central America. These flocks may contain 20 or more Golden-cheeked Warblers and a mix of other Neotropical migrants such as Wilson's Warblers, Blue-headed Vireos, and Black-and-White Warblers, as well as resident avian species.

Researchers estimate that only 28 percent of juvenile Golden-cheeked Warblers survive to breed the next year, but annual survival increases to land and wildlife and prefer woodlands over open pastures.

On the winter range, the Alliance for the Conservation of the Pine-Oak Forest in Mesoamerica has developed a regional conservation plan to safeguard this ecosystem as habitat for the Golden-cheeked Warbler. The plan proposes sustainable forestry practices that allow utilization of the forests while maintaining the species composition and structure required by the Golden-cheeked Warbler.

FEDERAL DOCUMENTATION:

Listing: 1990. Federal Register 55:18844–18845 (Emergency rule)

1990. Federal Register 55:53153–53160 (Final rule)

Critical Habitat: Not designated

Recovery Plan: US Fish and Wildlife Service. 1992. Golden-cheeked warbler (*Dendroica chrysoparia*) recovery plan. Albuquerque, NM. 88 pp.

Red-cockaded Woodpecker,
Picoides borealis

ROBERT ALLEN

FEDERAL STATUS: Endangered
TEXAS STATUS: Endangered

The Red-cockaded Woodpecker was first described in 1807 by the French naturalist Louis Jean Pierre Vieillot. Unaware of Vieillot's description, Alexander Wilson also described the species in 1810 and provided the common name still used today. A tuft of red feathers between an adult male's black crown and white cheek patch is the basis for the term "red-cockaded." The red tuft is displayed only for a brief period when a male is courting a female and remains hidden for most of the year. Throughout its range, there are 6,408 potential breeding groups of Red-cockaded Woodpeckers. Approximately 476 potential breeding groups occur in Texas, most of which occupy public lands scattered throughout the eastern portion of the state. Only 7 percent of the state's Red-cockaded Woodpeckers currently occur on private lands.

DESCRIPTION. Red-cockaded Woodpeckers are relatively small—adults are 8 to 9 inches long and weigh 1.5 to 1.75 ounces. Size varies geographically; larger birds are found in the northern latitudes. A conspicuous white cheek patch distinguishes the Red-cockaded Woodpecker from all other woodpecker species in its range (fig. 9.20). The top and back of the head are black, transitioning to black-and-white barring (a "ladderback") on the back and wings. The breast and belly are white to grayish white, and distinctive black spots are present along the sides of the breast, enlarging to bars on the flanks. Central tail feathers are black, whereas outer tail feathers are white with black barring. Adults have a black crown with a narrow white line above the black eye. The white cheek patch is separated from the white throat by a heavy black stripe. The bill is black and the legs are dark gray to black. Adults of both sexes have similar plumage and are generally indistinguishable in the field. Juveniles can be distinguished in the field until the first fall molt; juvenile males possess a red crown patch.

Figure 9.20. A Red-cockaded Woodpecker with provisions to feed its young. Photograph by James D. Childress.

DISTRIBUTION. The Red-cockaded Woodpecker is endemic to the open pine forests of the southeastern United States. Its historical range spans a large region extending from New Jersey, Maryland, and Virginia to Florida, west to Texas and north to portions of Oklahoma, Missouri, Tennessee, and Kentucky. The present range has contracted to Virginia, North Carolina, South Carolina, Georgia, Florida, Alabama, Mississippi, Arkansas, Louisiana, Oklahoma, and Texas. Within East Texas it is found in mature pine forests at scattered locations extending westward from the Sabine River to Cherokee, Montgomery, and Walker Counties.

HABITAT. Red-cockaded Woodpeckers utilize mature open pine and pine-hardwood forests for nesting and foraging habitat. Key aspects of nesting habitat include large pines greater than 60 years of age, absent or sparse midstory, herbaceous ground cover, and few canopy hardwoods, if any (fig. 9.21). Foraging habitat includes the habitat described above as well as open pine and pine-hardwood stands as young as 30 years of age. Fire is paramount to the development and maintenance of suitable habitat. Within Texas, Red-cockaded Woodpeckers utilize forests dominated by Longleaf, Shortleaf, Loblolly, and nonnative Slash Pines. Most suitable habitat is currently on US Forest Service lands, Texas A&M Forest Service lands, and a few private properties.

NATURAL HISTORY. Red-cockaded Woodpeckers are nonmigratory and territorial. They exhibit a complex social system described as a cooperative breeding system. Individuals live in a family group known as a potential breeding group, which comprises a breeding pair with up to four male offspring (rarely female) from previous years. These male offspring ("helpers") assist in raising the current year's offspring and defending the territory. The territory, which is termed a cluster, comprises multiple roost trees that are close to one another. Red-cockaded Woodpeckers nest and roost within cavities excavated in living pine trees, a trait unique among North American woodpeckers. This behavior may have evolved in response to conditions within a pyric ecosystem,

Figure 9.21. These marked Shortleaf and Longleaf Pines indicate that they are part of a Red-cockaded Woodpecker cluster. Photograph by Brian R. Chapman.

where standing dead pines would be consumed by fire. In addition to the cavity, the birds inflict wounds (resin wells) to the bole above and below the cavity. Sticky resin trickles down the bole and deters predators.

Breeding activity begins in April and ends in July. Renesting may occur if the initial nesting attempt is unsuccessful. Some pairs may raise a second brood in a single season, but double brooding is rare. Females typically lay three to five eggs in the breeding male's roost cavity. Eggs hatch after an incubation period of 10–12 days, and nestlings fledge from the nest cavity 20–27 days later. Fledglings may continue to be fed for several months, after which most female fledglings disperse to fill breeding vacancies elsewhere. Fledgling males either disperse or remain on their natal territory as helpers. Breeding activity ceases in July.

The prey base of Red-cockaded Woodpeckers consists mostly of eggs, larvae, and adults of ants, beetles, and roaches, various other insects, and spiders. Fruits and seeds are a minor portion of the overall diet. Larger, older trees are preferred for foraging. Males typically forage on the upper trunk and limbs, while females forage on the trunk below the crown. Foraging methods include flaking away bark to expose prey as well as probing crevices in the bark with the tongue.

REASONS FOR POPULATION DECLINE AND THREATS. Once common throughout the Southeast, the Red-cockaded Woodpecker initially declined because of intensive logging and land conversion throughout its range. By the early twentieth century, the original old-growth open pine forests in East Texas had been reduced to a few small isolated remnants. Later, fire suppression and detrimental silvicultural practices including clear-cutting, short rotation, and conversion to suboptimal pine species exacerbated the decline. By the 1990s only a few stable populations remained, and some, especially smaller populations, continue to decline. Increased knowledge of Red-cockaded Woodpecker population ecology and advancements in management techniques such as artificial cavity construction, translocation, and prescribed burning have been significant in stemming the decline.

Current threats to the species include fire suppression and the resultant intrusion of hardwoods; habitat fragmentation and its effects on genetic variation, dispersal, and demography; and the risks to small populations from random demographic, environmental, genetic, and catastrophic events.

RECOVERY AND CONSERVATION. The recovery goal for the Red-cockaded Woodpecker is to attain species viability by maintaining multiple large populations distributed so as to minimize threats from negative demographic, environmental, genetic, and catastrophic events. Recovery is focused on restoring open pine and pine hardwood forests of sufficient age and structure to sustain populations, with an emphasis on frequent prescribed fire and population monitoring.

Recovery criteria include establishing 11 primary core populations (10 containing at least 350

potential breeding groups and 1 with at least 1,000 potential breeding groups), 9 secondary core populations each containing 250 potential breeding groups, and 250 potential breeding groups spread throughout the southeastern United States. Primary core populations in Texas include the Sam Houston, Angelina, and Sabine National Forests, while the Davy Crockett National Forest serves as a secondary core population.

Multiple conservation actions are necessary to maintain and expand current populations within Texas. Creating the open forest structure required by the species may involve thinning mature pine and pine-hardwood stands to 40–60 square feet of basal area and removing midstory vegetation by various methods (mechanical removal, prescribed fire, or herbicides). The use of artificial cavities to supplement existing Red-cockaded Woodpecker clusters and to establish new clusters will continue to be vital for population stability and expansion until recovery is achieved.

FEDERAL DOCUMENTATION:
Listing: 1970. Federal Register 35:16047
Five-Year Review: 2005. Federal Register 70:53807–53808
Recovery Plan: US Fish and Wildlife Service. 2003. Recovery plan for the Red-cockaded Woodpecker (*Picoides borealis*): Second revision. Atlanta, GA. 296 pp.

Red-crowned Parrot,
Amazona viridigenalis

KARL S. BERG

FEDERAL STATUS: Candidate
TEXAS STATUS: Not Listed

Residents in the Lower Rio Grande Valley of Texas are sometimes jolted awake in the early morning by raucous screeches as flocks of Red-crowned Parrots journey noisily to feeding areas. These birds are likely descendants of vagrants that wandered from the lowlands of northeastern Mexico decades ago and established a resident population in South Texas. However, centuries ago there was likely suitable habitat along the lower Rio Grande, including the now largely decimated Texas Sabal Palm forests and Texas Ebony evergreen thorn forests, which may have supported populations of Red-crowned Parrots. In any event, the native population in Mexico continues to be highly threatened, but the population in Texas and many feral populations of released birds or their descendants elsewhere may be stable or increasing.

DESCRIPTION. As in many medium-sized parrots (13 inches in length), the body plumage of the Red-crowned Parrot is a bright iridescent green. A red crown accentuated posteriorly by a narrow fringe of violet-blue feathers tops the head (fig. 9.22). The red on the crown of the female does not extend as far back, and even less so on juveniles. Green feathers tipped with black give the hind neck a scaly appearance. A bare, whitish ring surrounds the eye, the beak is pale yellow, and yellowish-green feathers appear at the tip of the tail. A conspicuous red wing bar on the primary flight feathers is visible mainly in flight, but it is partially exposed while birds are perched. The iris is yellow in adults and black or grayish in juveniles.

DISTRIBUTION. The native range of Red-crowned Parrots encompasses about 19,000 square miles along the Gulf Coast lowlands of Mexico in Tamaulipas, San Luis de Potosí, and Veracruz. Largish flocks began appearing in the Rio Grande Valley of South Texas in the late 1980s and have been gradually increasing. Red-crowned Parrots in South Texas and extreme northeastern Mexico may have arrived as escaped pets, vagrants from farther south, or a combination. Feral populations exist in Puerto Rico, the Yucatán and parts of central Mexico, Hawaii, southern California, and Florida.

HABITAT. Within the southern portion of their native range in Mexico, Red-crowned Parrots inhabit tropical forests with trees that are 50 to 70 feet tall. In the northern, and drier, parts of their Mexican range, the parrots frequent Tamaulipan thornscrub habitats and taller vegetation in ravines and riparian zones. In urban settings, the parrots concentrate their activities in older neighborhoods and parks where large trees that provide food and nesting sites are abundant.

Figure 9.22. Red-crowned Parrot at the entrance to its nest cavity. Photograph by Karl S. Berg.

NATURAL HISTORY. Red-crowned Parrots forage opportunistically on ripened fruits, seeds, flower buds, flowers, and acorns. Although the birds often obtain food items from the crowns of trees, they also feed in low-lying shrubs. The parrots obtain water from fruit pulp and rarely drink.

Mature birds form lifelong pair bonds during the fall and winter months when the parrots congregate in large flocks. As flocks begin to dissipate in February, pairs begin defending previously used nests in existing tree cavities or locate new holes. Pairs perform loud duets at or around nest sites. The basic theme is six notes, three by the female and three by the male, given in rapid alternating succession and often ending in a "bugle" type sound by the male; the female gives the first note of the duet. Contact calls often given in flight can be harsh, guttural calls interspersed with higher-pitched whistles or screeches. Red-crowned Parrots do not excavate their own nest cavities, preferring instead to use natural holes or those hollowed out by woodpeckers. In South Texas, the parrots often use holes in dead palms for nesting sites. Nests of several pairs are sometimes near one another. As older palm snags can have multiple cavities, Red-crowned Parrots compete or breed concurrently in nest cavities with other species of parrots and cavity-nesting birds. Nesting can also occur in artificial structures in urban areas. Pairs can be seen copulating around nest sites around the time egg laying begins, in mid-March to early April.

Females lay eggs at approximately 48-hour intervals until the clutch of three or four eggs is complete. Females begin incubation after laying the first egg, and the young begin hatching asynchronously 27 days later at intervals reflecting the laying sequence. At hatching, the young are naked and their eyelids are fused, but they are soon covered by whitish down. Females brood the young for an average of 17 days and feed them twice daily. During incubation and brooding, males feed females twice daily; afterward both parents abandon the brood during most of the day, feeding nestlings in the morning and late evening. Nestlings begin to leave the nest in the order of hatching in late June to early July when they are about 53 days of age, but they remain near the nest, where the parents feed them for another three to five weeks. Parents continue to feed offspring for up to six months after fledging. The life span of wild birds is unknown, but captive individuals frequently live 20 years.

REASONS FOR POPULATION DECLINE AND THREATS. The IUCN lists the Red-crowned Parrot as endangered in northeastern Mexico. Local extirpations and overall population declines of the Red-crowned Parrot in many areas of Mexico have resulted from the combination of poaching for the pet trade and habitat destruction. Some poachers obtain nestlings by cutting down nest trees, thus reducing nest-site availability. Clearing of land for agriculture and lumber harvests also eliminates nesting and feeding habitats.

RECOVERY AND CONSERVATION. State, federal, and university partners conduct quarterly surveys of the Red-crowned Parrot population in South Texas. The Texas Parks and Wildlife Department (TPWD) has organized a working group to address conservation and protection of the species and hosts annual workshops to enlist support and train volunteers. The University of Texas Rio Grande Valley (UTRGV) partnered with the TPWD to erect 20 artificial nest cavities in greater Brownsville in 2016. The UTRGV also estimates breeding density and overall population sizes by conducting monthly roost counts in Brownsville. Similar efforts farther inland are underway by scientists from Texas A&M University. Two privately funded organizations, the American Bird Conservancy and the Rio Grande Joint Venture, are planning to develop a monitoring and conservation plan with the goal of protecting the bird's remaining critical habitat in Mexico.

FEDERAL DOCUMENTATION:
Listing: 2011. Federal Register 76:62016–62034
Critical Habitat: Not designated
Recovery Plan: Not developed

We can't live on this planet with a dead ocean.
— PAUL WATSON

The ocean basin known as the Gulf of Mexico is connected to the Atlantic Ocean by a narrow channel, the Florida Straits, between Florida and Cuba and to the Caribbean Sea by the Yucatán Channel between Cuba and Mexico. The Gulf is 913 miles wide and covers an expanse of 615,000 square miles. The midportion is deep, but about half of the basin consists of waters above a relatively shallow coastal shelf. Many of the Gulf of Mexico species listed as threatened or endangered enter the Gulf as wanderers from populations normally found in the Atlantic Ocean. Accounts describing the protected species in the Gulf of Mexico are provided in this chapter.

Caribbean Electric Ray,
Narcine bancroftii

PHILIP MATICH

FEDERAL STATUS: Candidate
TEXAS STATUS: Not Listed

DESCRIPTION. The Caribbean Electric Ray is a relatively small batoid (a superorder of cartilaginous fishes that includes rays, skates, and sawfishes), reaching up to approximately 2 feet in length. Its circular, flattened body is brown to dark orange with dark patches and a white or light gray underside (fig. 10.1). The ray has a moderately long snout, and it can extend the small mouth on the underside of its body to capture prey. Two dorsal fins of equal size are attached to a short, thick tail with a prominent triangular caudal fin lacking a spine. The ray's most prominent feature is a pair of electric organs extending from the front of the eyes to the rear of the body. These organs have generated considerable interest among researchers studying electric rays.

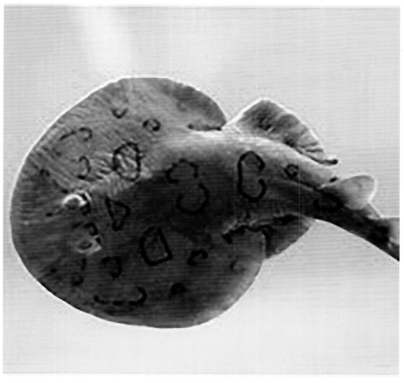

Figure 10.1. Caribbean Electric Ray. Photograph by Wayne Hoggard, Southeast Fisheries Science Center, National Oceanic and Atmospheric Administration.

DISTRIBUTION. The Caribbean Electric Ray inhabits the coastal waters of the western Atlantic, ranging from North Carolina through the Gulf of Mexico and the Caribbean Sea, reaching the northern coast of South America.

HABITAT. A bottom dweller, the Caribbean Electric Ray lives in shallow coastal waters including coral reefs, bays, and river mouths ranging from intertidal waters to about 100 feet. The rays often bury themselves in soft sandy sediment for protection and may use seagrass beds for foraging.

NATURAL HISTORY. The Caribbean Electric Ray utilizes shallow habitats to forage on worms, small fishes, crustaceans, and other invertebrates while avoiding large bony fishes and sharks that pose threats as predators. The ray is nocturnal, using its electric organs to stun prey while foraging at night, and burying itself in the sediment during the day. The electric organ produces a shock exceeding 30 volts, and when threatened, the ray surprises predators with a brief pulse to deter or temporarily stun them. It is unclear whether the strength and efficiency of electric organs develop through ontogeny, but neonates are born with the ability to emit electric pulses.

Throughout most of its range, the Caribbean Electric Ray uses shallow inshore waters during summer months, but as waters cool in the fall and winter, it migrates to deeper waters to thermoregulate. It is unclear whether these migrations may also be driven by reproductive behavior or foraging ecology.

Female Caribbean Electric Rays are more abundant than males, which may mitigate vulnerability to population declines attributed to slow swimming speeds and limited capacity to escape extended predatory attacks. Females typically mature at larger sizes than males and become sexually mature in as little as two years. After mating, the gestation period is less than one year, often much shorter. Caribbean Electric Rays are ovoviviparous (fertilized eggs hatch within the body and exit as live young). There is no placental attachment between the mother and embryo, but yolks and secretions from the uterus provide energy and nutrients to developing pups. Litter sizes range from 4 to 20 offspring.

REASONS FOR POPULATION DECLINE AND THREATS. Human actions pose the greatest threat to Caribbean Electric Rays. Although the rays are not internationally harvested, their slow swimming speeds limit their ability to escape trawls through bycatch reduction devices, and they are frequently captured as bycatch in commercial shrimp trawling and artisanal seine-net fishing. While rays are often discarded upon capture, survival rates are believed to be low after release—pregnant females often abort embryos. While the global population status is unknown, estimates from bycatch records suggest declines exceeding 95 percent at the end of the twentieth century. Recent surveys of batoid communities suggest that the ray's population is low.

In addition to fishing pressure, habitat degradation also poses a threat to Caribbean Electric Rays. Because the rays have small activity areas, coastal development, pollution, and removal of potential prey organisms may have heightened detrimental impacts on them compared to more mobile, wide-ranging species. Although declines have been documented only through fishing bycatch data in the United States, unregulated fishing in the Caribbean has likely resulted in similar declines. The ray's status in Texas waters is unknown, but disturbances including dredging and trawling likely have negative impacts on survival rates.

RECOVERY AND CONSERVATION. Despite a rapid and dramatic decline in abundance during the last 20 years, no specific conservation measures protect the Caribbean Electric Ray. Unfortunately, data are limited; thus, the first step toward conserving this critically endangered ray should be gathering data on population size and structure, survival rates, trends in population status, and behavioral ecology to identify critical habitat and prey species.

Developing a management strategy for the Caribbean Electric Ray is challenging. Fisheries do not target this species, and the rays are often released. Bycatch reduction devices that target sea turtles and other charismatic marine taxa may help reduce fishery-induced mortality, but they are likely ineffective because of the sluggish nature of the ray. More stringent local fishing regulations during mating and pupping seasons may aid in population recovery, and strict release-upon-capture regulations will also aid conservation efforts. Fortunately, unlike other many other elasmobranchs (a subclass of cartilaginous fishes that includes sharks, skates, rays, and sawfishes), the Caribbean Electric Ray becomes sexually mature relatively quickly, and some populations are heavily skewed in favor of females. Thus, population recovery may be more rapid than that of other elasmobranchs if anthropogenic impacts can be mitigated.

FEDERAL DOCUMENTATION:
Listing: 2014. Federal Register 79:4877–4883
Critical Habitat: Not designated
Recovery Plan: Not developed

Sawfishes

PHILIP MATICH

Smalltooth Sawfish, *Pristis pectinata*
Largetooth Sawfish, *Pristis pristis*

Bearing elongated rostrums laterally studded with sharp teeth used for foraging and defense, sawfishes are among the most morphologically distinct and recognizable fishes in coastal ecosystems. Sawfishes are also among the largest marine dwellers, with several species exceeding 23 feet in length and weighing more than 1.1 tons. Historically, sawfishes frequented Texas coastal waters and the inshore bays, estuaries, and rivers along the entire Gulf of Mexico. Use of shallow coastal habitats placed them in proximity to humans and likely led to local extirpations within Texas waters from exploitation, entanglement in fishing nets, and habitat degradation. Similar worldwide patterns of population decline and extirpations have rendered sawfishes among the world's most endangered marine species.

DESCRIPTION. Sawfishes are unique among cartilaginous fishes. Their most distinguishing feature, the long, toothed rostrum, makes up more than 20 percent of the body length. Extending forward from the head, it is lined with enlarged dermal denticles (small, tooth-shaped scales) and electrosensory pores. Their elongated bodies, strong trunks, and caudal fins are similar to those of sharks, and their flattened body shape and skeletal structure resemble those of rays (fig. 10.2). Sawfishes are typically dark gray to blackish brown dorsally, paler along the margins of the fins, and grayish white to pale yellow on the ventral side. The teeth inside the mouth are small and rounded.

DISTRIBUTION. Historically, sawfishes occupied shallow, tropical and subtropical coastal waters and rivers in 90 countries and territories

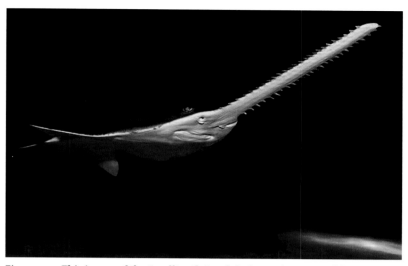

Figure 10.2. This image of the Smalltooth Sawfish shows the external features of sawfishes, especially the long, tooth-studded rostrum. Photograph by D. Doss Robertson, Smithsonian Institution.

worldwide. Sawfishes once occupied all Texas coastal waters and shallow waters in the Gulf of Mexico, but they may now be restricted to only the Gulf waters of Florida and neighboring islands in the Caribbean Sea.

HABITAT. Sawfishes typically inhabit coastal habitats less than 33 feet deep, especially environments with sandy and muddy bottoms, mangrove shorelines, and seagrass beds. Shallow waters may serve as a defense against sharks and crocodilians and aid in foraging for fishes and crustaceans. Juveniles typically occupy water less than 3.5 feet deep. Although some individuals undertake long-distance migrations along coastlines, site fidelity to small activity areas is typical.

NATURAL HISTORY. Sawfishes frequently remain in shallow nearshore waters, where they use their rostrum to disturb bottom-dwelling fishes and crustaceans and slash through schools of small-bodied fishes. Prey are swallowed whole. Rostrum teeth are not replaced when chipped or broken but continue to grow. In some areas sawfish movements appear tidally coordinated, likely allowing individuals to remain within certain water depths to exploit foraging opportunities, avoid predators, and linger in warm waters to accelerate metabolism and growth rates. Social behavior is unknown among sawfishes, but aggregations of 300 or more have been caught in nets, suggesting that schooling occasionally occurs.

Sawfishes bear live young—eggs are retained in the uterus of females, where embryos receive nourishment from yolk sacs. Up to 20 progeny per female are born in brackish or fresh waters. At birth the young are between 24 and 36 inches in total length, and a sheath encases their soft, leathery rostrums. Juvenile sawfishes grow rapidly, especially during their first year, but adult sawfishes exhibit slow growth compared to bony fishes. Sawfishes often exhibit age-related habitat shifts; juveniles use freshwater and brackish habitats in rivers and estuaries for the first few years of life before increasing their use of marine waters as they mature. Some species may live more than 50 years in undisturbed habitats.

REASONS FOR POPULATION DECLINE AND THREATS. Conservation efforts targeting sawfishes have only recently been enacted despite declines approaching 99 percent of historical population numbers. Most sawfishes are taken as bycatch—they are often lethally entangled in nets. Consequently, population declines have been largely overlooked or disregarded by management agencies. Sawfishes exhibit life history characteristics, such as late age at maturity and low fecundity, that constrain population recovery in the absence of targeted conservation actions.

Exploitation, unintended mortality as bycatch, and habitat loss continue to be among the most significant threats to sawfishes worldwide. Sawfishes are still targeted in some areas for their meat, fins, and rostrums, especially in countries lacking the infrastructure to enforce conservation laws. The most precipitous population declines, however, likely result from fishing bycatch. Gill net bans in the United States have reduced sawfish bycatch, but other fishing gear (trawls, seines, and lines) targeting bony fishes or sharks also poses threats. Population estimates across the world suggest that harvesting sawfishes is not sustainable and will ultimately lead to extirpation or extinction.

Habitat loss and degradation also contribute to population declines. Mangrove shorelines and coastal estuaries serve as critical habitats for sawfishes, especially juveniles, and continued alteration of these habitats is a key conservation concern for coastal marine ecological communities. Within the United States, Florida serves as the last stronghold for sawfishes. While conservation measures are designed to increase habitat availability, considerable management challenges persist in regions where residents and tourists impact coastal waters.

RECOVERY AND CONSERVATION. All species of sawfishes are listed in Appendix I of the Convention on International Trade in Endangered Species of Wild Fauna and Flora (CITES), which bans commercial international trade in sawfishes or their parts. Although all sawfish species are endangered, they are legally protected in only 16 of the 90 countries and territories that historically provided sawfish habitat. The countries that protect sawfishes, however, include about 80 percent of current sawfish habitat within their waters (including the United States). Many countries ban shark finning, requiring instead that shark fins remain attached through landing, a constraint that may improve sawfish monitoring efforts and reduce illegal fishing. Regional fishing restrictions have also improved management and conservation of sawfishes and other batoids.

Only one species of sawfish, the Smalltooth Sawfish, currently resides within US waters. The Largetooth Sawfish remains on the endangered species list despite its extirpation from US waters. Although critical habitat is not designated for the Smalltooth Sawfish, the recovery plan urges protection of mangrove-lined estuaries in Florida to safeguard nursery habitats and facilitate growth of the adult population. Outside the United States, Brazil, Mexico, Nicaragua, Mauritania, and Guinea-Bissau all have legislation that protects sawfishes in the Atlantic Ocean.

Smalltooth Sawfish, *Pristis pectinata*
FEDERAL STATUS: Endangered
TEXAS STATUS: Endangered

DISTINGUISHING FEATURES. As the name suggests, the Smalltooth Sawfish has smaller and more numerous (20–34) rostral teeth than the Largetooth Sawfish. The caudal (tail) fin of the Smalltooth Sawfish does not have a lower lobe, and the rostrum constitutes about 25 percent of the sawfish's body length. The second dorsal fin is nearly as large as the first, and the origin of the first dorsal fin is above the origin of the pelvic fins.

SPECIFIC DISTRIBUTION AND HABITAT. The Smalltooth Sawfish was historically found in tropical and subtropical marine and estuarine waters of the Atlantic, but the species has experienced an 81 percent decline in geographic range. In the western Atlantic, Smalltooth Sawfish ranged from New York to the Gulf of Mexico and throughout much of the Caribbean Sea. The species has been extirpated outside of Florida in the United States and eastward beyond the Turks and Caicos Islands. Its current range in the eastern Atlantic is now limited to Sierra Leone.

Smalltooth Sawfish frequently inhabit shallow coastal waters and exhibit a preference for warm waters (> 69.8°F). Pregnant females give birth in estuarine habitats, and juveniles use shallow estuaries, nearshore sand and mud banks, mangrove shorelines, and seagrass beds as nurseries for their first few years of life. Juveniles often seek mangrove prop roots for foraging and refuge.

FEDERAL DOCUMENTATION:
Listing: 2005. Federal Register 70:69464–69466
Critical Habitat: Not designated
Recovery Plan: National Marine Fisheries Service.
    2009. Smalltooth Sawfish recovery plan (*Pristis pectinata*). Silver Spring, MD. 102 pp.

Largetooth Sawfish, *Pristis pristis*
FEDERAL STATUS: Endangered
TEXAS STATUS: Not Listed

DISTINGUISHING FEATURES. Largetooth Sawfish have much larger, but many fewer (< 24) rostral teeth than the Smalltooth Sawfish. The rostrum constitutes approximately 20 percent of the body length (fig. 10.3). The caudal fin has a distinct lower lobe, and the second dorsal fin is smaller than the first.

Figure 10.3. Largetooth Sawfish. Photograph by J. Patrick Fischer, Creative Commons.

SPECIFIC DISTRIBUTION AND HABITAT. The Largetooth Sawfish was more widely distributed than the Smalltooth Sawfish, but its geographic range has been reduced by 61 percent. It may be extirpated from 28–55 of the 75 countries where it historically occurred. Largetooth Sawfish were found in many tropical and subtropical coastal ecosystems, including the Gulf of Mexico. They no longer exist along the Gulf Coast. Within the United States, Largetooth Sawfish were most frequently found on the Texas coast, predominantly in the Galveston-Freeport area. Their historical range in the western Atlantic Ocean extended from Florida to Uruguay. The Brazilian-Amazonian region serves as one of the last western Atlantic refuges for this species, but it also occurs throughout the Indian and Pacific Oceans.

FEDERAL DOCUMENTATION:
Listing: 2015. Federal Register 80:3914–3916
Critical Habitat: Not designated
Recovery Plan: Not developed

Hammerhead Sharks

PHILIP MATICH

Scalloped Hammerhead, *Sphyrna lewini*
Great Hammerhead, *Sphyrna mokarran*

Large-bodied and highly mobile, hammerhead sharks are top predators in Texas coastal waters. They play significant roles in regulating fish populations (including other elasmobranchs) through consumptive interactions (sharks eat fishes) and indirect interactions that alter species traits and behaviors (sharks scare smaller fishes). While some shark species are more commonly encountered by fishermen, hammerhead sharks roam both shallow and deeper waters of the Texas Gulf Coast from South Padre Island to the Bolivar Peninsula and Sabine Pass.

DESCRIPTION. The cephalofoil (hammerhead) of hammerhead sharks distinguishes them from all other shark species. Dorsoventrally flattened and laterally extended, the head possesses an array of electroreceptors (the ampullae of Lorenzini). Eyes are located on the sides of the cephalofoil for optimal prey detection. Hammerheads have tall first dorsal fins and gray, brown, or olive-green bodies countershaded by white undersides that provide camouflage in the ocean. Their razor-sharp teeth are designed to handle large prey, including stingrays and smaller sharks. Hammerhead sharks exhibit a range of sizes, from Scalloped Bonnetheads (maximum length of 3 feet) to Great Hammerheads (up to 20 feet in length), and morphological characteristics (head and fin shape, coloration) that can change as sharks grow.

DISTRIBUTION. Hammerheads are distributed globally in coastal waters along continental and insular shelves; some occupy deeper waters near shelves, but they are not considered oceanic species. The sharks engage in long-distance migrations and have seasonal ranges in both tropical and cold regions. Hammerheads frequent coastal ecosystems throughout the Gulf of Mexico and Caribbean, including offshore waters along the entire Texas coastline.

HABITAT. Hammerheads occupy many habitats—estuaries, bays, coral reefs, deeper waters, and (rarely) freshwater rivers. Like many shark species, hammerheads spend a large proportion of their time near the substrate but make daily vertical migrations for foraging. Small juveniles segregate from adults to avoid cannibalism.

NATURAL HISTORY. Hammerheads exhibit species-specific differences in population structure and behavior. Some species (Scalloped Hammerhead, Smooth Hammerhead) form large schools with hundreds of individuals, often with segregation between males and females. Others (Great Hammerhead) are more solitary and exhibit less social behavior, especially as adults. Whether solitary or schooling, many species undertake long seasonal migrations, which may allow them to maintain homeostasis, track schooling fishes and other prey species, or engage in reproduction. Females of some species make inshore migrations to give birth in shallow waters after gestation periods of less than one year.

Hammerheads are viviparous (bear live young); a yolk-sac placenta provides energy and nutrients to embryos. After birth, neonates rely on energy reserves in body tissues to avoid starvation while developing foraging skills. Juvenile hammerheads use shallow coastal waters early in their lives to avoid predators, such as larger sharks, and feed on bony fishes, crabs, shrimps, squids, and other invertebrates. Some species shift their diets and habitat use as they increase in size, transitioning from smaller-bodied prey in shallow waters to larger-bodied prey, including other sharks and stingrays, in deeper waters. Some hammerhead species remain mesopredators (medium-sized predators) in their respective ecosystems where larger sharks pose a threat of predation throughout their lifetimes. Other species reach sizes allowing them to escape hazards from other large sharks—the risks they face are primarily human induced.

REASONS FOR POPULATION DECLINE AND THREATS. Despite increased conservation efforts, many shark species continue to decline across the world, a cause of considerable concern because their life histories make them sensitive to changes in population size and structure. Many sharks, including hammerheads, exhibit low ecological resiliency because their slow growth rates, late age at maturity, and low fecundity lead to inherently low population growth rates.

Among the most pressing threats to hammerheads and other sharks is fishing pressure. Sharks are targeted for their meat, livers, fins, and gill rakers (toothlike structures protecting their delicate gills). Hammerhead sharks have relatively large fin-to-body-size ratios, making them prized by the shark fin trade. Schooling behavior makes some hammerheads particularly vulnerable to fisheries. Even when not targeted, sharks are frequently caught as bycatch in commercial fisheries for large-bodied marine fishes such as tuna or billfish. Because they exhaust themselves while being caught, hammerheads exhibit significant postrelease mortality.

Habitat degradation is also of concern for many coastal shark species, including hammerheads, because they rely on inshore waters for foraging and nursery habitats for juveniles. Habitat disturbance and pollution reduce habitat quality and availability as well as food sources for coastal and oceanic sharks. Fishing pressure places sharks directly at risk through mortality and threatens their survival by removing their prey from the environment. In Texas, juvenile hammerhead sharks use coastal ecosystems during their early years; however, continued human-caused depletion of natural resources and ecological disturbance may limit hammerhead growth and survival rates, impacting adult populations.

Human use of shark habitats for personal enjoyment (swimming, fishing, and surfing) has also led to declines in shark populations through lethal management practices. Despite detrimental impacts to ecosystems and the low frequency of shark encounters by humans, some countries have implemented protocols limiting shark access to highly used coastal waters. The removal of sharks found close to highly populated areas mimics other successful, but ecologically harmful, predator population control programs developed in the past for wolves, mountain lions, and bears.

RECOVERY AND CONSERVATION. All hammerhead species except the Bonnethead Shark are currently experiencing population declines. Despite the need for conservation actions, the population dynamics and natural history of most hammerhead species are poorly known. Across many geographic regions, it is difficult to establish accurate population estimates and delineate critical habitats. Research and monitoring efforts have increased, but the methods used are not uniform, making it challenging to compare findings across studies.

Conservation and management practices in some countries may be offset by fishing pressures and habitat degradation in other locales because many hammerhead species migrate across international boundaries. Shark populations are not monitored globally, and underreported or unreliable landings data in many countries make it difficult to assess temporal shifts in species landings among fisheries.

Despite these challenges, many governments are improving their management strategies for sharks within their respective borders by creating sanctuaries and marine protected areas and by more rigorously managing shark fishing. Many countries have also banned shark finning by requiring that shark fins remain attached through landing. Such restrictions may improve shark monitoring efforts and mitigate illegal fishing. Within the United States, only Florida has outlawed fishing for hammerheads. Texas allows only one shark per day to be taken by a fisherman, and there are size limitations protecting small juveniles that most frequently reside in shallow, vulnerable habitats. Pregnant females, however, may be at risk in these habitats when giving birth.

Scalloped Hammerhead, *Sphyrna lewini*
FEDERAL STATUS: Threatened
TEXAS STATUS: Not Listed

DISTINGUISHING FEATURES. Hammerheads are often distinguished by species-specific differences in their cephalofoil. Scalloped Hammerheads have an indentation in the center of the anterior margin of their curved head, and

Figure 10.4.
Scalloped
Hammerhead.
Photograph by
Barry Peters,
Creative
Commons.

Figure 10.5.
A school of
Scalloped
Hammerheads.
Photograph by
Seawatch.org,
Wikimedia
Commons.

of South America, the Caribbean Sea, and Gulf of Mexico and have been detected along the East Coast of the United States beyond Maine into Nova Scotian waters.

Scalloped Hammerheads occur in the waters of continental and insular shelves and adjacent deep waters, including inshore waters of bays and estuaries, where juveniles are typically found. As Scalloped Hammerheads grow, usage of deeper waters generally increases. The sharks thermoregulate using deeper, colder waters during summer months and shallower, warmer waters above the thermocline in winter. Scalloped Hammerheads often gather in schools segregated by sex (fig. 10.5). Females use inshore waters to pup but typically frequent pelagic habitats earlier than males.

FEDERAL DOCUMENTATION:
Listing: 2014. Federal Register 79:38214–38242
Critical Habitat: 2015. Federal Register 80:71774–71784 (none in the United States)
Recovery Plan: Not developed

Great Hammerhead, *Sphyrna mokarran*
FEDERAL STATUS: Not Listed
TEXAS STATUS: Not Listed
IUCN STATUS: Endangered

DISTINGUISHING FEATURES. Great Hammerheads are the largest species of hammerhead and in Texas may attain weights of 871 pounds and lengths of 13.7 feet. The nearly straight front margin of an adult's head has a shallow notch in the center (fig. 10.6). Great Hammerheads also possess strongly serrated teeth and distinctly concave pelvic fins with curved rear margins.

SPECIFIC DISTRIBUTION AND HABITAT. Found throughout the Gulf of Mexico and its inshore waters, Great Hammerheads also occur worldwide in tropical, subtropical, and some temperate waters. Their latitudinal range is more restricted than that of Scalloped Hammerheads. Great Hammerheads frequent inshore and offshore habitats, including coral reefs, lagoons, and deeper

two smaller lateral indentations, giving the head a "scalloped" appearance (fig. 10.4). Scalloped Hammerheads have a moderately slender body and smooth-edged to weakly serrated teeth.

SPECIFIC DISTRIBUTION AND HABITAT. Scalloped Hammerheads inhabit coastal waters in tropical, subtropical, and some temperate regions worldwide. In the northwestern Atlantic, Scalloped Hammerheads are found in the equatorial waters

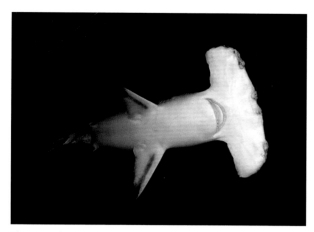

Figure 10.6. Great Hammerhead. Photograph by Jim Capaldi, Wikimedia Commons.

waters. Individuals are typically solitary. They migrate latitudinally poleward during summer months and toward the equator in winter months to thermoregulate.

FEDERAL DOCUMENTATION:

Listing: Endangered, IUCN

Critical Habitat: Not described; IUCN includes the Gulf of Mexico

Recovery Plan: Not developed

Sea Turtles

DONNA M. SHAVER
AND MARY M. STREICH

Loggerhead Sea Turtle, *Caretta caretta*
Green Sea Turtle, *Chelonia mydas*
Leatherback Sea Turtle, *Dermochelys coriacea*
Hawksbill Sea Turtle, *Eretmochelys imbricata*
Kemp's Ridley Sea Turtle, *Lepidochelys kempii*

The Gulf of Mexico and the Texas coast support five of the world's seven sea turtle species. The five species—Loggerhead, Green, Leatherback, Hawksbill, and Kemp's Ridley Sea Turtles—are all classified as threatened or endangered under the US Endangered Species Act. The most abundant sea turtles in Texas waters are the Green, Kemp's Ridley, and Loggerhead. All five species nest on Texas beaches, but Kemp's Ridley is the most frequently documented nesting species.

DESCRIPTION. Sea turtles, also called marine turtles, are air-breathing reptiles that are well adapted to the marine environment. Common characteristics of the species include streamlined bodies, relatively nonretractile extremities, extensively roofed skulls, and large paddle-like limbs. All are large—adult body weights vary from 75 to 2,000 pounds—and all have substantial salt glands to excrete excess salt ingested with seawater and food. Four of the five species are classified in the family Cheloniidae, the hard-shelled turtles, which possess a carapace (top shell) composed of bony plates covered with horny scutes. The Leatherback Sea Turtle, which is classified in the family Dermochelyidae, has a flexible carapace with a nearly continuous layer of small dermal bones lying just below an outer layer of leathery skin.

DISTRIBUTION. Sea turtles are found in temperate and tropical seas throughout the world. Distribution of the five species occurring in Texas marine waters varies seasonally and geographically, and by species and life stage. Sea turtles tend to be more most abundant in waters less than 160 feet deep and often frequent bays, estuaries, and passes, which some species use as developmental habitat. Passes also serve as areas of transit between Texas bays and the Gulf of Mexico. Kemp's Ridleys predominate along the upper Texas coast, whereas Green Sea Turtles are more common in shallower southern Texas waters. Although sea turtles live most of their lives in the marine environment, adult females return to beaches to lay and bury their eggs on land. Nesting occurs on Gulf of Mexico beaches statewide but is more common along the lower coast, particularly at Padre Island National Seashore (PINS), the only location in Texas where all five sea turtle species come ashore to nest.

HABITAT. Habitats used differ with species and life stage. Sea turtles have divergent and often specialized feeding habits, and these influence habitat use. For example, Kemp's Ridley Sea Turtles forage mostly on crabs and seek bay areas with muddy and sandy bottoms, whereas bay areas with seagrass bottoms are preferred by Green Sea

Figure 10.7. Green Sea Turtles often graze on seagrasses in shallow Texas bays and lagoons. Photograph by P. Lindgren, Wikimedia Commons.

Figure 10.8. During an *arribada* in 2011, hundreds of female Kemp's Ridley Sea Turtles swarmed ashore to lay eggs in the warm beach sands near Rancho Nuevo, Mexico. Photograph by Toni Torres, National Marine Fisheries Service and Gladys Porter Zoo.

Turtles that forage mostly on seagrasses and algae (fig. 10.7). Jettied passes, piers, pilings, and oil and gas platforms provide structure for attachment of algae and invertebrates and attract other marine life, which is exploited by other foraging sea turtles. Gulf waters provide migratory, internesting, foraging, and hatchling dispersal habitat. Sandy Gulf beaches provide nesting sites.

NATURAL HISTORY. Adult sea turtles migrate between adult foraging grounds and nesting beaches. Mature males and females mate along the migratory pathway to the nesting beach or off the nesting beach during the spring. Females

nest every one to four years, and during those years they emerge onto a beach to deposit one to eight clutches, each containing about 100 eggs. Kemp's Ridleys nest mostly during daylight hours, usually on windy days, and often in *arribadas* (synchronous arrivals; fig. 10.8). In contrast, other species nest singly and mostly at night. Eggs incubate for about two months—sex of the hatchlings is determined by incubation temperature.

After emerging from the nest, hatchlings scramble over the beach to the surf. Once in the water, they swim offshore. After the swim frenzy period, posthatchlings (oceanic-stage juveniles) spend their early life near the surface in pelagic waters of the Gulf of Mexico or Atlantic Ocean, taking shelter in floating material such as seaweed or drift lines of floating debris. Leatherbacks remain pelagic through their juvenile and adult life stages, but after the first few years of life, juveniles of other species return to nearshore areas or move into bays and estuaries, where they eat and grow. Sea turtles reach maturity between 10 and 50 years of age, depending on the species.

REASONS FOR POPULATION DECLINE AND THREATS. Populations of all five sea turtle species have declined because of human-related and natural factors. Incidental capture (bycatch) by commercial and recreational fisheries remains a significant threat (fig. 10.9, left). Commercial harvest for eggs, meat, and other products (skin, scutes, oil) has also caused population declines but has been prohibited in the United States since 1978 and in Mexico since 1990. Large-scale collection of eggs from Rancho Nuevo, Mexico, the primary nesting beach for Kemp's Ridley Sea Turtles, coupled with incidental capture of juveniles and adults decimated the numbers of that species by the early 1960s. By 1900, the Green Sea Turtle population inhabiting Texas waters had crashed from the commercial overharvest of turtles and from hypothermic stunning of turtles during severe freezes in the 1890s. Hypothermic stunning continues to be a significant threat to Green Sea Turtles in Texas. Sea turtles in the marine environment are also susceptible to degradation of foraging habitat, pollution and toxins, hazardous

Figure 10.9. Still entrapped in a shrimp net, a dead Kemp's Ridley Sea Turtle drifted ashore on Padre Island (left). Turtle Excluder Devices installed on shrimp nets allow sea turtles like this young Loggerhead to escape (right). Photographs courtesy of the National Park Service (left) and the National Marine Fisheries Service (right).

materials, entanglement in and ingestion of marine debris, vessel strikes, dredging, power plant entrapment, oil and gas development, oil spills, increased human presence and harassment, poaching (illegal harvest, primarily for eggs and meat), diseases (including fibropapillomatosis), and parasites. In addition to these population stressors, nesting turtles, eggs, and hatchlings are also impacted by loss and degradation of nesting habitat through beach erosion, development, coastal armoring, artificial lighting, beach nourishment, beach maintenance and cleaning, beach vehicular traffic, items discarded on the beach, holes dug and ruts made in the sand, predation by native and nonnative predators, climate change, and thermal stress. Egg and hatchling survival are also reduced by high tides and inundation, heavy rains, and entanglement in large amounts of seaweed (*Sargassum* spp.) that often accumulate on the beach and in the surf along the Texas coast.

RECOVERY AND CONSERVATION. The five sea turtle species are protected under the US Endangered Species Act of 1973. Since 1977, the National Oceanic and Atmospheric Administration (NOAA) Fisheries and the US Fish and Wildlife Service have shared jurisdiction for recovery and conservation of sea turtles listed under the Endangered Species Act. NOAA has jurisdiction for sea turtles while they are in marine environments, and the US Fish and Wildlife Service has jurisdiction while they are on land. Biologists from these agencies, along with subject matter experts, have developed recovery plans for each species. These plans outline threats to the species and actions to guide recovery efforts.

Fisheries interactions pose significant threats to sea turtles, and many efforts have been made to reduce incidental capture. Sea turtles are protected under the Magnuson-Stevens Fishery Conservation and Management Act. Since the early 1990s, NOAA has implemented other at-sea conservation measures for sea turtles including Turtle Excluder Device (TED) regulations for trawl fisheries (fig. 10.9, right), large circle hooks in longline fisheries, and time and area closures for fisheries. The Texas Parks and Wildlife Department, which also has jurisdiction over sea turtles in Texas, has also instituted time and area closures for fisheries, which have benefited sea turtles.

Sea turtles are migratory and travel to multiple jurisdictions during their life. Most sea turtles inhabiting Texas waters likely hatched in other states and countries, particularly Mexico. Thus, recovery of these species in Texas depends on conservation actions undertaken on all sea turtle nesting beaches and marine habitats. In 1990, a ban on taking any species of sea turtle was effected by Mexican presidential decree. Active conservation programs are ongoing in Mexico and the United States, and research and public education efforts are being conducted throughout the range of these species.

Texas participates in the Sea Turtle Stranding and Salvage Network, which has documented sea turtles found stranded (washed ashore, alive or dead) in the United States since 1980. These data provide minimum estimates of mortality—not all dead turtles wash ashore. Live turtles are taken to rehabilitation facilities, where they receive medical treatment. Many sea turtles brought to rehabilitation facilities in Texas survive and are repatriated to Texas waters. The stranding network also generates data useful for studying trends in these species and identifying threats to sea turtles in the marine environment.

The largest sea turtle conservation project in Texas has been the binational effort to form a secondary nesting colony of Kemp's Ridley Sea Turtles at PINS. PINS was selected as the location for this project because the National Park Service can safeguard the nesting habitat, turtles, and eggs on the longest stretch of undeveloped barrier island beach in the United States. From 1978 to 1988, 22,507 Kemp's Ridley eggs were collected at the primary Kemp's Ridley nesting beach at Rancho Nuevo in Tamaulipas, Mexico, and brought to PINS for incubation. The hope was that packing the eggs in Padre Island sand and releasing the hatchlings on the beach at PINS would imprint the turtles to Padre Island and cause them to return at adulthood to nest. After reaching the surf zone, the hatchlings were captured and transported to the NOAA Fisheries Laboratory in Galveston, where they were reared in captivity for 9 to 11 months. This "head-starting" continues, allowing the turtles to grow large enough to be tagged for future recognition and avoid most predators after release. Surviving "Padre Island imprinted" yearlings are released at a variety of locations, but mostly off North Padre and Mustang Islands.

The numbers of sea turtle nests found in Texas generally increased from 1980 through 2016, but some nests were likely undetected. Some Padre Island and Mexico imprinted head-started turtles have nested in Texas, but most nesting turtles have been from wild stock. A secondary nesting colony consisting of about 55 percent of the Kemp's Ridley Sea Turtles nesting in the United States continues to develop at PINS.

None of the sea turtle species occurring in Texas are nearing recovery and delisting. Although the juvenile Green Sea Turtle population has increased in Texas and Green Sea Turtles are now found stranded more frequently than any other sea turtle species, the population inhabiting Texas waters likely remains below former levels. Kemp's Ridley recovery efforts were showing promising signs of success from the 1990s through 2009, but exponential increases in nest numbers ceased in 2010, and concern about the status of the species was renewed. Numbers of Loggerhead, Hawksbill, and Leatherback turtles occurring in Texas appear to be stable. Because all five sea turtle species occurring in Texas are long-lived and migratory, recovery efforts for them must be long term and cooperative with other entities throughout their range.

Loggerhead Sea Turtle, *Caretta caretta*
FEDERAL STATUS: Threatened
TEXAS STATUS: Threatened

DISTINGUISHING FEATURES. Loggerhead Sea Turtles have five lateral scutes on each side of the carapace, three inframarginal scutes without pores on the plastron (bottom shell), and more than one pair of prefrontal scales (between the eyes and nostrils) on their comparatively large head (fig. 10.10). Adults are reddish brown above and pale yellow below. Hatchlings are reddish brown above and tan to reddish brown below. Adult carapace length reaches 36 inches, and weight nears 250 pounds.

SPECIFIC DISTRIBUTION AND HABITAT. In Texas, Loggerheads occur primarily in nearshore Gulf of Mexico waters and occasionally within passes and bays. They eat mostly crabs, mollusks, and other invertebrates and are often found in association with oil and gas platforms. From 1980 through 2016, 87 nests were recorded in Texas (range 0–13 nests per year), including 63 on North

Figure 10.10. Loggerhead Sea Turtle. Photograph by Bachrach44, Wikimedia Commons.

Figure 10.11. Green Sea Turtle. Photograph by Michael Lusk, US Fish and Wildlife Service.

Padre Island (mostly at PINS). Nesting has been recorded from May through mid-August.

FEDERAL DOCUMENTATION:

Listing: 1978. Federal Register 43:32800–32811

Critical Habitat: 2014. Federal Register 79:39855–39912

Recovery Plan: National Marine Fisheries Service and US Fish and Wildlife Service. 2008. Recovery plan for the northwest Atlantic population of the Loggerhead Sea Turtle (*Caretta caretta*), second revision. Silver Spring, MD. 325 pp.

Green Sea Turtle, *Chelonia mydas*

FEDERAL STATUS: Threatened

TEXAS STATUS: Threatened

DISTINGUISHING FEATURES. Green Sea Turtles have four lateral scutes on each side of the carapace, four inframarginal scutes without pores on the plastron, one pair of prefrontal scales on the head, and a serrated lower jaw (fig. 10.11). Adults are mottled brown above and pale yellow below. Hatchlings are black above and white below, with white edges on their carapace and flippers. Adult carapace length is 36–48 inches, and weight is 300–400 pounds.

SPECIFIC DISTRIBUTION AND HABITAT. South Texas bays, estuaries, and passes serve as important developmental habitat for juvenile

Green Sea Turtles, which feed mostly on seagrasses and algae. Between 1980 and 2016, 66 nests were recorded in Texas (range 0–15 nests per year), including 60 on North Padre Island (mostly at PINS) and 6 on South Padre Island. Nesting has been recorded from June through mid-September.

FEDERAL DOCUMENTATION:

Listing: 1978. Federal Register 43:32800–32811

Critical Habitat: 1998. Federal Register 63:46693–46701

Recovery Plan: National Marine Fisheries Service and US Fish and Wildlife Service. 1991. Recovery plan for U.S. population of Atlantic Green Sea Turtle. Washington, DC. 52 pp.

Leatherback Sea Turtle, *Dermochelys coriacea*

FEDERAL STATUS: Endangered

TEXAS STATUS: Endangered

DISTINGUISHING FEATURES. The carapace of Leatherback Sea Turtles is composed of leathery, oil-saturated connective tissue overlying a nearly continuous layer of small dermal bones. Seven prominent keels (longitudinal ridges) taper to a blunt point posteriorly on the carapace (fig. 10.12). They have scaleless, leathery skin. Hatchlings and adults are primarily black with some pinkish-white spotting. Hatchlings have white striping on the seven ridges and on the edges of their flippers. The

Figure 10.12. Leatherback Sea Turtle. Photograph courtesy of the US Fish and Wildlife Service.

Leatherback is the largest living sea turtle. Adult carapace length is up to 78 inches, and weight approaches 2,000 pounds.

SPECIFIC DISTRIBUTION AND HABITAT. Leatherbacks are the most pelagic and globally distributed sea turtle. They prefer deeper Gulf of Mexico waters where they forage primarily on jellyfish, and they rarely occur in inshore waters of Texas passes and bays. Historical records indicate that Leatherbacks nested in Texas during the 1920s and 1930s at sites later included in PINS. Only one recent nest has been recorded in Texas, in June 2008 at PINS.

FEDERAL DOCUMENTATION:
Listing: 1970. Federal Register 35:8491–8498
Critical Habitat: 1979. Federal Register 44:17710–17712
Recovery Plan: National Marine Fisheries Service and US Fish and Wildlife Service. 1992. Recovery plan for Leatherback Turtles in the U.S., Caribbean, Atlantic and Gulf of Mexico. Washington, DC. 65 pp.

Hawksbill Sea Turtle,
Eretmochelys imbricata
FEDERAL STATUS: Endangered
TEXAS STATUS: Endangered

DISTINGUISHING FEATURES. Hawksbill Sea Turtles have four lateral scutes on each side of the carapace, four inframarginal scutes without pores on the plastron, and two pairs of prefrontal scales on the head (fig. 10.13). The back is serrated with overlapping scutes, and the elongated head tapers to a point at the beak-like mouth. The carapace of juveniles and adults is dark brown with streaks of yellow, orange, red, or black, and the plastron is yellow. Hatchlings are dark brown. Adult carapace length is 25–35 inches, and weight is 100–150 pounds.

SPECIFIC DISTRIBUTION AND HABITAT. Hawksbills are the most tropical of all sea turtles. In Texas, they occur in nearshore waters of the Gulf of Mexico and waters within passes armored with jetties, but rarely in bays. They prefer coral reef and rocky habitats where they forage primarily on sponges and wedge in crevices for resting. Only one Hawksbill nest has been recorded in Texas. It was found in June 1998 within the boundaries of PINS.

FEDERAL DOCUMENTATION:
Listing: 1970. Federal Register 35:8491–8498
Critical Habitat: 1998. Federal Register 63:46693–46701

Figure 10.13. The hook-like beak is a distinguishing feature of the Hawksbill Sea Turtle. Photograph courtesy of the National Oceanic and Atmospheric Administration.

Recovery Plan: National Marine Fisheries Service and US Fish and Wildlife Service. 1993. Recovery plan for Hawksbill Turtles in the U.S., Caribbean Sea, Atlantic Ocean, and Gulf of Mexico. St. Petersburg, FL. 52 pp.

Kemp's Ridley Sea Turtle,
Lepidochelys kempii
FEDERAL STATUS: Endangered
TEXAS STATUS: Endangered

DISTINGUISHING FEATURES. The Kemp's Ridley is the smallest living sea turtle. The adult carapace length is 24–28 inches, and weight is 75–100 pounds. Kemp's Ridley Sea Turtles have five lateral scutes on each side of the carapace, four inframarginal scutes with pores on the plastron, and more than one pair of prefrontal scales on their comparatively large head (fig. 10.14, left). The adult carapace is often wider than long. Adults are olive green to gray above and pale yellow below. Hatchlings are charcoal gray above and below. Juveniles have keeled carapaces that are serrated along the trailing edge (fig. 10.14, right).

SPECIFIC DISTRIBUTION AND HABITAT. In Texas, Kemp's Ridley Sea Turtles occur in nearshore Gulf of Mexico waters, bays, and passes, where they feed mostly on crabs, and occasionally on fishes and molluscs. Upper and mid-Texas

coastal bays and passes with muddy and sandy bottoms serve as important developmental habitat for juveniles. In Texas, nesting has been recorded from April through mid-July.

FEDERAL DOCUMENTATION:
Listing: 1970. Federal Register 35:18319–18322
Critical Habitat: Not designated
Recovery Plan: National Marine Fisheries Service, US Fish and Wildlife Service, and SEMARNAT. 2011. Bi-national recovery plan for the Kemp's Ridley Sea Turtle (*Lepidochelys kempii*), second revision. Silver Spring, MD. 156 pp.

Baleen Whales

BERND WÜRSIG

Fin Whale, *Balaenoptera physalus*
Humpback Whale, *Megaptera novaeangliae*

Fin and Humpback Whales are not common in the Gulf of Mexico. Although rarely sighted at sea, they occasionally strand themselves on shore. They are likely to be vagrants from the North Atlantic, and no estimates of population for either species in the Gulf are available. Bryde's Whale is the only baleen whale commonly encountered in the Gulf of Mexico.

DESCRIPTION. Fin and Humpback Whales have plates of baleen embedded in the gums where teeth occur in most mammals. Baleen is composed of keratin (like fingernails and hair) and hangs from the roof of the mouth to serve as a

Figure 10.14. An adult Kemp's Ridley Sea Turtle (left). Hatching Kemp's Ridleys scramble seaward after leaving the nest (right). Photographs courtesy of the US Fish and Wildlife Service (left) and the National Park Service (right).

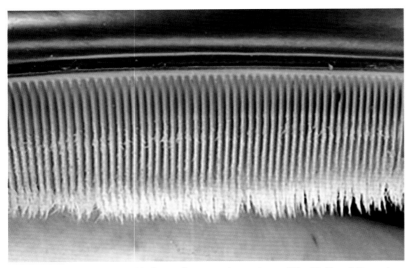

Figure 10.15. An example of baleen. Photograph courtesy of the National Oceanic and Atmospheric Administration.

filtering mechanism during foraging. Both species are members of a group commonly known as "rorquals," a term referring to the throat grooves that expand when the whales lunge forward into schools of krill (euphausiid crustaceans) or fishes. After each lunge, rorquals incompletely close their mouths and exert muscular action to contract the throat pleats. The resulting pressure squirts water through the baleen plates (fig. 10.15) and out of the mouth, allowing the plates to capture and compact a prey bolus, which the whale swallows. Lunge feeding is a highly energetic activity—the operation of the lower jaw musculature during lunge feeding by the Blue Whale and Fin Whale, the world's two largest animals, is deemed the greatest single biomechanical function on Earth.

DISTRIBUTION. Both species of baleen whale occur in all major oceans except extreme polar regions, and both migrate long distances to feed at high latitudes in summer and to lower latitudes closer to the equator for breeding and calving in winter. Humpback Whales engage in one of the longest seasonal migrations of any mammal on Earth. They generally occur in shallower waters than Fin Whales and often breed near coastal areas or around islands in tropical waters. These species rarely enter the Gulf of Mexico as vagrants from the North Atlantic.

HABITAT. Fin and Humpback Whales frequent both shallow and offshore open ocean waters. They often feed in productive cooler or cold waters where krill or fishes concentrate. In some areas, Humpback Whales tend to feed on near-bottom-dwelling schools of fishes, such as the Sandlance. While both species feed on both krill and fishes, Fin Whale diets include more krill than fishes, especially in higher latitudes of the Southern Hemisphere.

NATURAL HISTORY. Mating in all rorquals occurs in tropical waters during the winter. Gestation lasts slightly less than one year, so a single calf tends to be born during the winter. Lactation lasts about six to seven months in Fin Whales, but close to one year in Humpback Whales. Both species become sexually mature before age 10 but continue to grow throughout life, albeit slowly after sexual maturity. Interbirth intervals in females tend to be two to three years, and longevity is unknown, although recognized individuals of both species over 30 years old have been recorded.

Male Humpback Whales produce complicated songs during the breeding season, but how these songs help them acquire suitable mates remains unknown. Females may preferentially select a mate from males that produce the appropriate song at a certain intensity. Male Fin Whales and possibly all rorquals sing basic songs, which may also play a role in sexual selection.

REASONS FOR POPULATION DECLINE AND THREATS. The largest single reason for past declines of most rorquals, including Fin and Humpback Whales, is modern (twentieth-century) intensive hunting from factory ships. Use of explosive harpoons, devices to inject gases into a whale carcass to keep it floating while being rendered into oil and meat for human and animal consumption, and modern freezing methods improved hunting efficiency and profit margins. Because they bred and calved close to tropical islands and other shores in winter, Humpback Whales were hunted intensively in the early 1900s.

Since the decline of whaling in the 1970s and early 1980s, rorquals have generally increased in numbers. Some rorqual whaling by Japan in the

Antarctic and by small island nations in tropical parts of the world still occurs. Fishing entanglement also causes some whale mortality, especially in nearshore areas where nets, crab pot lines, and other devices are commonly employed. Occasional hunting and fishing entanglement do not presently appear to represent major problems to populations, however. A growing threat is boat strikes by large, high-speed ferries and ocean-going cargo ships where shipping lanes and rorqual habitat coincide. Depletion of fish stocks from overfishing in many parts of the world may be an especially significant problem in some enclosed waters.

There is no reason to believe that any rorquals were ever numerous in the Gulf of Mexico. The Gulf does not have appropriate habitats for feeding or breeding by Fin and Humpback Whales. When North Atlantic populations were larger, before modern whaling practices were employed, more vagrants likely entered the Gulf than do so now.

RECOVERY AND CONSERVATION. Rorquals and all marine mammals are now better protected by most countries than they were the mid-twentieth century. The US Marine Mammal Protection Act of 1972 (and the many similar protections in British Commonwealth and European countries) and multiple-country mandates by the International Whaling Commission were instrumental in strongly curtailing whaling. These laws regulate trade in whale products such as the meat, oil, baleen, and teeth of Sperm Whales. They also protect marine mammals against fishery-related problems and noise from tourism vessels, navy sonar sounds, and nearshore habitat degradation. None of these mandates and efforts are perfect, and constant vigilance over environmental regulation and enforcement is necessary.

Fin Whale, *Balaenoptera physalus*
FEDERAL STATUS: Endangered
TEXAS STATUS: Endangered

DISTINGUISHING FEATURES. Adult male Fin Whales can reach 85 feet in length, but females tend to be 5–10 percent larger than males at the same age. Although generally dark above and light below,

the color pattern on Fin Whales is asymmetrical—much of the lower jaw and baleen plates on the right side is white, while the left side is much darker or black. Fin Whales have a sleek, streamlined body, a V-shaped head (viewed from above), a prominent dorsal fin, and pectoral fins (flippers) that do not appear excessively large (fig. 10.16). Their dorsal fin, however, is larger than that of the largest rorqual, the Blue Whale. The Fin Whale's streamlined body allows fast swimming speeds, which prompted the moniker "greyhound of the seas."

SPECIFIC DISTRIBUTION AND HABITAT. Fin Whales occupy offshore habitats throughout the world. Most occupy productive feeding habitats in higher-latitude waters during the summer months before migrating to clear tropical waters in the winter to breed. During southward migrations, Fin Whales in the western Atlantic Ocean move rapidly from the waters of Labrador and Newfoundland to the West Indies. Vagrant Fin Whales occasionally enter the Gulf of Mexico, and some approach the Texas coast.

FEDERAL DOCUMENTATION:
Listing: 1970. Federal Register 35:8491–8498
Critical Habitat: Not designated
Recovery Plan: National Marine Fisheries Service. 2010. Final recovery plan for the Fin Whale (*Balaenoptera physalus*). Silver Springs, MD. 121 pp.

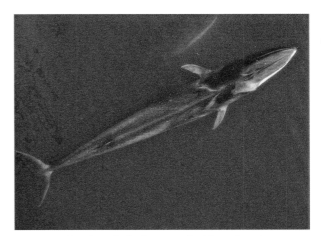

Figure 10.16. A Fin Whale photographed from the air. Photograph courtesy of the National Oceanic and Atmospheric Administration.

Humpback Whale, *Megaptera novaeangliae*

FEDERAL STATUS: Delisted in 2016
TEXAS STATUS: Endangered

DISTINGUISHING FEATURES. Humpback Whales are large-bodied whales with a maximum length that seldom exceeds 50 feet. Their body is dark above and light below. Their small dorsal fin is set upon a "hump" on the back, and a set of dorsal knobs is visible behind the dorsal fin (fig. 10.17). Humpback Whales have the longest flippers of any whale or dolphin species (the generic name means "giant wing). Humpback Whales have numerous dermal tubercles (knobby structures called "stove bolts" by whalers) on the middle head ridge and along other parts of the head, chin, and lower jaw.

SPECIFIC DISTRIBUTION AND HABITAT. Humpback Whales live in all major oceans. Individuals that infrequently visit the Gulf of Mexico, including Texas waters, are likely vagrants from the West Indies distinct population segment. The National Marine Fisheries Service recently delisted this Humpback Whale population (from endangered status) because of recovery. Whales in the West Indies distinct population segment feed in the Gulf of Maine, along the east coast of Canada, and off the west coast of Greenland. During the winter, they breed and calve in the Atlantic Ocean from Cuba to northern Venezuela.

FEDERAL DELISTING DOCUMENTATION: 2016. Federal Register 81:93639–93641

Figure 10.17. Humpback Whale. Photograph courtesy of the National Oceanic and Atmospheric Administration.

Sperm Whale, *Physeter macrocephalus*

BERND WÜRSIG

FEDERAL STATUS: Endangered
TEXAS STATUS: Endangered

The Sperm Whale is the largest of the toothed whales (odontocetes) and is the largest toothed animal on Earth. This deep-diving species feeds largely, but not exclusively, on squid.

DESCRIPTION. Sperm Whales are gray above and light below, with a pronounced hump where the dorsal fin is located (fig. 10.18). They are sexually dimorphic—males are larger than females, averaging about 49 feet long and about 80,000 pounds as adults. Females average about 36 feet in length and weigh about 44,000 pounds. The heads of males grow disproportionately larger in relation to their body size—the heads of mature older males may be almost one-third their body length—whereas young male and adult female heads are about one-fourth their entire length. This is the only whale to have a single blowhole, which is asymmetrically located on the left side of the head. Sperm Whales have large, rounded teeth in the lower jaw and only sockets in the upper jaw to house the teeth when the mouth is closed.

DISTRIBUTION. Sperm Whales inhabit deep waters throughout the world's oceans. In the Gulf of Mexico, they are especially common south of the Texas-Louisiana Shelf and in the Mississippi and De Soto Canyons to the east. Their population in the Gulf is largely, but not completely, separate from the Sperm Whales in North Atlantic tropical waters. Seasonal migration in the Gulf of Mexico is limited, if it occurs at all.

HABITAT. Water depths greater than 1,960 feet where sea surface temperatures are 60°F or higher provide optimal habitats for Sperm Whales. The whales rarely visit areas where water is less than 1,960 feet deep. Most Sperm Whales remain far away from land, where the continental shelf in the Gulf of Mexico descends to appropriate depths.

Figure 10.18. A female Sperm Whale with her calf. Photograph by Gabriel Barathieu, Creative Commons.

NATURAL HISTORY. Sperm Whales feed on medium-sized to large squid and bottom-dwelling (demersal) and middepth (mesopelagic) fishes. Because they are deep divers and can access food at depths greater than 2,600 feet, they do not rely on nighttime feeding (many odontocete cetaceans feed nocturnally when squid and fishes of the deep scattering layer come closer to the surface).

Females become sexually mature at about age 9–10 and thereafter bear a single calf about every 5 years. The sexual displays and mating behavior of Sperm Whales have not been well described. The gestation period lasts 14–16 months. The young nurse for several years but can ingest solid food after their first year. Young males remain with their mothers and other females until they become sexually mature, but females often stay within the matriarchal society for life. Pods (groups) of about 12 related females travel together and care for each other's young. Males leave the pod at around age 10 and tend to cluster with other young males as they mature. These clusters break up with age, and fully mature adult males (generally older than 25 years) wander alone. Females become reproductively senescent (enter menopause) at about age 40, and both males and females can reach ages of at least 50 years.

Sperm Whales produce clicks for echolocation and codas (click trains) for communication. The clicks reverberate in the large head and develop different tonal qualities related to the size of the head, the age of the individual, and the dialects in different ocean regions. These tonal qualities may also serve as elements of individual recognition, social signaling, and sexual displays. The disproportionately large head of adult male Sperm Whales may therefore serve as both a secondary sexual characteristic (like the large antlers of older deer) and a long-distance sound-mediated sexual characteristic (male clicks differ and may imply information about individual vitality).

Sperm Whales occasionally mass strand, with up to many dozens of animals on a beach at a time. The cause of mass stranding remains unknown, but the occurrences may be related to navigation mistakes near a shelving shoreline and the social integrity of pods that prevents some animals from leaving others.

REASONS FOR POPULATION DECLINE AND THREATS. Sperm Whales, including those in the Gulf of Mexico, were hunted intensively for oil in the 1800s. After World War II, they were hunted primarily for meat until a worldwide moratorium on whaling occurred in 1988. Populations, especially large males, were decimated in all oceans of the world, but Sperm Whales fared better than other large (baleen) whales. In the North

Atlantic, Sperm Whale populations now appear to be doing well. Not only are they not hunted, but deep-living squid and fishes are not the focus of worldwide fisheries, so their prey has been relatively unaffected.

Sperm Whales are occasionally killed by entanglement in fishing gear, by ship strikes, and possibly by contaminants in their bodies. They also appear to be particularly vulnerable to high-intensity sound. It is unknown whether large-scale use of loud navy sonar, for example, may be causing them to surface too rapidly and die of embolisms.

RECOVERY AND CONSERVATION. The worldwide Sperm Whale population is not well documented, but estimates tend to range from the hundreds of thousands to 1.5 million. For the western North Atlantic, the most recent (2015) stock assessment is about 2,288 animals. This stock cannot be differentiated from that of the Gulf of Mexico, where the best recent estimate is 763 Sperm Whales. The species remains listed as endangered because many populations have not fully recovered from the effects of past hunting. International trade in Sperm Whale products, including teeth (for ornamental scrimshaw and other decorative uses), is prohibited under CITES.

Sperm Whales have become popular subjects for ecotourism (whale watching) in New Zealand, the Azores, Dominica, and Norway. Although whale watching in the Gulf has not yet turned into a regular commercial activity, some excursions occur out of ports in southern Louisiana, where Sperm Whales come within 30 miles of land.

West Indian Manatee,
Trichechus manatus

BRIAN R. CHAPMAN

FEDERAL STATUS: Threatened
TEXAS STATUS: Endangered

The historical ranges of the two West Indian Manatee subspecies overlapped along the Texas coast. Although few specimen records exist, sightings of manatees in Texas waters occur sporadically when individuals stray west from Florida or north from Mexico. Decades often pass between sightings, but Trinity Bay near Galveston and Cove Harbor near Rockport have hosted manatees during the past 10 years.

DESCRIPTION. West Indian Manatees are massive marine mammals with flipper-like forelimbs, no hind limbs, and a horizontally flattened tail (fig. 10.19). Their sparsely haired skin is typically dark grayish brown (some are black) and rubbery. The nostrils, which are on the upper snout, open when the animal surfaces and close when it dives. An inner membrane draws across the eyes to protect them when a manatee is submerged. Tiny ear openings are present, but pinnae are lacking. Short, stiff bristles surround the wide, flattened lips. Adults average about 10 feet in length and 2,200 pounds, but individuals may reach lengths of 25 feet and weights nearing 3,570 pounds.

DISTRIBUTION. Manatees occur primarily in Florida and southeastern Georgia. Individuals occasionally wander as far north on the Atlantic Coast as Rhode Island and as far west on the Gulf Coast as Texas. The subspecies that inhabits the Mexican Gulf Coast rarely ventures north of Veracruz.

HABITAT. West Indian Manatees inhabit freshwater, brackish, and marine systems along

Figure 10.19. West Indian Manatees. Photograph by Steve Hillibrand, US Fish and Wildlife Service.

coasts and riverine areas throughout their range. They usually remain near shorelines where they feed on submergent, emergent, and floating vegetation. When threatened, manatees escape to deep channels.

NATURAL HISTORY. Manatees communicate with high-pitched squeaks, and individuals or small groups often return to the same wintering or summering habitats for many years. Breeding can take place throughout the year. Males congregate around a female in estrus for many days while members of the breeding herd compete for access to the female of interest. After a 13-month gestation period, females usually give birth to a single calf, which remains with her for up to two years. Males reach sexual maturity by three to four years of age and females by age five. The life span may exceed 50 years.

Manatees feed opportunistically on a variety of aquatic plants, including seagrasses, marine algae, Water Hyacinth, Hydrilla, and Smooth Cordgrass. Sources of freshwater are often sought for drinking, and manatees sometimes gather to drink from water hoses at docks.

Their low metabolic rate requires manatees to seek warmer waters in the fall. Manatees that cannot find warm habitats during the winter are susceptible to cold stress and death. A West Indian Manatee rescued from Trinity Bay (Chambers County) in November 2014 displayed some signs of such stress.

REASONS FOR POPULATION DECLINE AND THREATS. Unregulated commercial and subsistence hunting in the 1800s likely reduced manatee populations throughout their range. The primary threats to the West Indian Manatee today include collisions with boats and boat propellers, habitat loss and fragmentation, and entanglement with fishing gear. Manatees also succumb to toxins released during harmful algal blooms (Red Tide), cold weather, tropical storms and hurricanes, tidal entrapments, and disease.

RECOVERY AND CONSERVATION. The West Indian Manatee was downlisted in 2017 from endangered to threatened, indicating a degree of recovery success. Establishment of restricted areas and minimum wake zones for boats in manatee habitats has reduced mortality rates in Florida. In 1991, an estimated 1,267 manatees inhabited Florida waters, but the population today exceeds 6,300.

FEDERAL DOCUMENTATION:
Listing: 2017. Federal Register 82:16668
Critical Habitat: 1976. Federal Register 41:41914
Recovery Plan: US Fish and Wildlife Service. 2001. Florida Manatee recovery plan (*Trichechus manatus latirostris*), third revision. Atlanta, GA. 144 pp.

PART THREE

A Glance at the Past

11 VANISHED: THE STATE'S LOST FAUNA

BRIAN R. CHAPMAN & WILLIAM I. LUTTERSCHMIDT

When we try to pick out anything by itself, we find it hitched to everything else in the Universe. — JOHN MUIR

Prior to 1700, the area that eventually became Texas was a wilderness containing extensive forests, prairies, deserts, coastal plains, and marshes. These pristine habitats, punctuated by thousands of clear springs feeding about 3,700 streams and 15 major rivers, were inhabited by abundant wildlife and numerous small groups of Native Americans. Most early human inhabitants were nomadic, camping for a season around springs or near rivers and subsisting by hunting and fishing before moving on; a few groups established more permanent lodging and developed farms where they grew maize, beans, pumpkins, and watermelons.

Although relations between tribes were not always peaceful, their impacts on land, water, and wildlife were localized and minimal.

In the eighteenth century, Spain began to establish Catholic missions in Texas. The main purpose of the missions was to convert the indigenous people, so the settlements that developed around the churches were small and primitive (fig. 11.1). Other than the priests and friars assigned to the missions, few European Americans ventured into the untamed region. Beginning in 1824, however, the Mexican government encouraged foreign settlers to purchase land in

Figure 11.1. Mission Concepción, established in 1716, was one of several built in the San Antonio area by the Spanish to serve the indigenous inhabitants of Tejas. Photograph by Travis Witt, Creative Commons.

the Tejas territory, and settlers began to immigrate there from Mexico. What had been only a trickle of immigrants from the 24 states representing the United States soon turned into a stream of colonists eager to take advantage of inexpensive land prices and tax deferrals offered by Mexico. Contrary to Mexico's prohibition of slavery in the territory, many brought slaves with them. Conflicts soon escalated between the government and the immigrants, who considered themselves "Texians" rather than Mexican nationals. Fearing loss of control in Tejas, Mexico banned further immigration from the United States in 1830. Tensions between the Texians and the government escalated and eventually led to the Texas Revolution.

When Texas became a state in 1845, the population hovered around 200,000. Soon thereafter, spurred by inexpensive land, abundant natural resources, and the expansion of railroads, the stream of immigrants became a flood. Within two decades, the rapidly escalating human population in the state approached 1 million. Most of these inhabitants were in the eastern half of the state, where farms proliferated and vast pine forests stimulated the growth of a timber industry.

Lured by the promise of jobs, workers and their families poured into the East Texas forests to harvest pines and hardwoods, produce charcoal, and refine turpentine and other naval stores. Timber companies constructed more than 200 railroads to serve 1,800 sawmills and nearly 3,700 other logging-industry sites. About 100 of the sawmills produced more than 80,000 board feet of lumber daily, and the wood was shipped throughout the world. Only a few lumber companies of that era practiced sustainable forestry. Most preferred instead to "cut and run" by moving their operations to virgin timberlands farther west after the pristine pine savannas were reduced to bleak graveyards of stumps and stubble. In abandoned cutover areas, erosion contributed loads of mud and silt to streams and rivers and impacted the survival of freshwater mussels and fish. Where the loud chattering calls of the Ivory-billed Woodpecker once echoed through majestic wooden columns, silence reigned (fig. 11.2). The

Figure 11.2. The Ivory-billed Woodpecker, a denizen of East Texas forests and swamps, was the largest woodpecker in North America. Painting by Mark Catesby (1682–1749) was published prior to 1923 and is in the public domain.

primary food of the nation's largest woodpecker — large wood-boring beetle larvae that inhabited the oldest pine snags — disappeared with the huge trees. With the absence of both food and nest trees, these magnificent birds rapidly declined. Thus, the first widespread human impacts on the flora and fauna of Texas were likely felt in the Piney Woods — but they would not be the last.

Decades before the last old-growth pines of East Texas were felled and carted away on railroad log cars, however, the original High Plains grazer, the American Bison, had been removed by what can only be described as "systematic slaughter." By the 1870s, fewer than 20 Bison, the last survivors of the southern herd that once numbered in the millions, remained in the state. Fortunately, these were saved from destruction by Molly (Mary

Ann) Goodnight, wife of famed rancher Charles Goodnight. Molly rescued calves orphaned by hunters, bottle-fed them, and nurtured the iconic beasts. From this small collection of waifs, the Goodnights eventually developed a substantial Bison herd. Descendants of this herd were reintroduced into Yellowstone National Park in 1902, while others became the Texas State Bison Herd at Caprock Canyons State Park (Briscoe County). Although Molly never received enough credit for her foresight, she should be remembered as the first Texan to prevent an extinction.

By the end of the nineteenth century, the human population of Texas topped 3 million. To supply the water needs of the burgeoning population and to irrigate crops in portions of the semiarid plains, thousands of wells were dug to tap underground aquifers. Soon, many springs that once sustained Native Americans and the European Americans that displaced them ceased to flow. Pools once nourished by springs with auspicious names such as Big Spring, Roaring Spring, and Silver Springs dried and their cracked bottoms filled with dust, detritus, and tumbleweeds. Untold numbers of tiny fishes and invertebrates, species that had washed into the ponds when glacial melt flooded the land, disappeared unnamed and as unknown as if they had never existed.

Despite its immense size, Texas originally contained just a single natural lake—Caddo Lake—in northeastern Texas. To provide a reliable source of water for humans, livestock, and irrigation, to drive gristmills and sawmills, and to control periodic floods, Texans elsewhere in the state began to construct dams on rivers. For example, by 1849 eight dams, including the Spring Lake Dam (still in existence), had been built on the upper San Marcos River. The early dams were crude by modern standards, not much more than simple barriers that created small reservoirs (fig. 11.3, top). Modern dams, which now impound over 200 large lakes in the state, are designed to release water periodically. The releases maintain downstream currents or mimic flood-pulse cycles, but many release water from the lake bottom (fig. 11.3, bottom). These flows of cold water often kill aquatic organisms,

Figure 11.3. Initial construction on Saffold Dam in Seguin began in the 1880s to impound a small lake as a reliable source of water to power a gristmill (top). Cold water released from the bottom of larger reservoirs often kills organisms and alters aquatic ecosystems far downstream (bottom). Photographs by Larry D. Moore, Creative Commons (top), and Bob McMillan, courtesy of the Federal Emergency Management Agency (bottom).

including fishes and freshwater mussels, and alter the aquatic ecosystems for many miles below the dam. Most dams block fish migrations, alter flood-pulse regimes, and alter sediment transport. The impoundments themselves eliminate stream habitats required by some aquatic organisms and have contributed to declines in the state's aquatic biodiversity.

There is no way to know how many animal species became extinct during the early days of Texas settlement. Although many large, obvious organisms that once inhabited the state were documented in the journals of early explorers and surveyors, the few biologists who ventured into the untamed area did not collect or catalog the tiny creatures that crawled between rocks, occupied leaf litter, or swam among algae in spring pools. The number of species endemic to only one or more springs in the state suggests that the number of species rendered extinct when various springs dried may have been enormous. With each cessation of spring flow, entire ecosystems, each consisting of many coexisting species, disappeared forever, leaving no trace of their existence.

Reductions in biodiversity continued into the twentieth century. The population of Texas continued to expand, reaching nearly 6 million by 1930, exceeding 11 million by 1970, and soaring well above 25 million by 2010. Urban populations, especially those in Houston and along the I-35 corridor, experienced the greatest growth. City boundaries expanded to accommodate densely packed housing, grocery stores, shopping malls, and pavement. Huge tracts of oak-juniper habitat surrounding San Antonio and Austin were cleared, eliminating habitats that once supported nesting Golden-cheeked Warblers and Black-capped Vireos, two species now listed as endangered. Greater demands were placed on water resources while aquifer recharge zones were simultaneously being hardened with concrete and macadam. Storm sewers, built to carry rainfall runoff from impervious urban surfaces, delivered oils, fertilizers, and other pollutants to aquifers, streams, and rivers. The addition of toxic chemicals to aquatic systems from both urban and agricultural runoff further affected the reproduction and survival of freshwater mussels and other sensitive aquatic organisms. The decline and loss of biodiversity, which accelerated during the last half of the twentieth century, appears to be a continuing trend.

Only a few of the small organisms now known to be extinct were documented before they vanished in the mid-1900s, often before biologists had time to learn much about them. Specimens of some can be examined in museum collections and a few are preserved for display in public exhibits, but most disappeared long ago. Modern citizens were denied an opportunity to observe them in the wild. What birder would not pay dearly to see an Ivory-billed Woodpecker? Sadly, the calls of an Ivory-billed Woodpecker, preserved in a single scratchy film recording made in 1935, provide the only hint of the sounds that echoed daily through the primeval East Texas forests over a century ago. The advance of civilization in Texas provides many opportunities for its citizens, but it has come with a cost—the decline and disappearance of native species.

Space limitations prevent a complete listing of the species whose presence once made Texas a much more interesting place. Brief descriptions—a mere sampling—of enigmatic species that inhabited Texas when the settlers arrived, but which are now extinct, are provided here. Each of these species was selected to demonstrate some of the many ways that a unique set of occurrences can lead to extinction.

ESTELLINE CRAB (*HEMIGRAPSUS ESTELLINENSIS*). Crabs typically exist only in oceanic environments, but the Estelline Crab persevered in Estelline Salt Spring ever since an ancient geologic period when seas covered the state (fig. 11.4). Located near Estelline (Hall County) at the northeastern base of the Texas Panhandle, the spring is over 500 miles from the Gulf of

Figure 11.4. The Yellow Shore Crab, which lives on the Pacific Coast, is the closest living relative of the Estelline Crab. Photograph by Ciar, Creative Commons.

Mexico. Sixteen specimens were collected at the salt spring in 1962 and more were observed, but the species was extinct by 1964. The disappearance coincided with the construction of a dam by the US Army Corps of Engineers. About 3,000 gallons of hypersaline water per minute fed a large pool that drained into the Prairie Dog Town Fork of the Red River. The dam prevented saltwater from flowing downstream, but evaporation from the surface increased the salinity of water in the pool and likely caused the unique crab's extinction. A barnacle that was also observed in the pool was never collected or identified.

AMISTAD GAMBUSIA (*GAMBUSIA AMISTADENSIS*).

The Amistad Gambusia was a small, guppy-like fish that inhabited a single locality in West Texas (fig. 11.5). Its habitat, Goodenough Spring (Val Verde County), was the state's third-largest natural spring. Goodenough Spring was submerged under 70 feet of water when the Amistad Reservoir filled in 1968. Prior to the completion of Amistad Dam on the Rio Grande, living specimens of Amistad Gambusia were collected and maintained at a University of Texas laboratory and at the Dexter National Fish Hatchery in New Mexico. Both captive populations were corrupted by hybridization with closely related mosquitofish. The Amistad Gambusia was declared extinct by the US Fish and Wildlife Service in 1987.

Figure 11.6. The San Marcos Gambusia (fig. 5.13), now believed to be extinct, once inhabited the crystal-clear waters of the San Marcos River. Photograph by WisdomFromIntrospect, Creative Commons.

SAN MARCOS GAMBUSIA (*GAMBUSIA GEORGEI*).

The cool, clear water flowing from San Marcos Springs provided a thermally stable habitat for the San Marcos Gambusia (see fig. 5.13). The small fish lived in Spring Lake and the headwaters of the San Marcos River in Hays County (fig. 11.6). Like other gambusia species, the San Marcos Gambusia gave birth to live young, but little else is known of its natural history. It was last collected in the wild in 1983 and has not been observed in the river or springs since then. Its disappearance has been attributed to reduced water flow from the springs and increased water pollution from the runoff of nearby cities. Nonnative fishes, which now inhabit the river, prey on smaller fishes such as the San Marcos Gambusia and compete with them for food. The tiny fish has not been declared extinct, but its continued existence is unlikely.

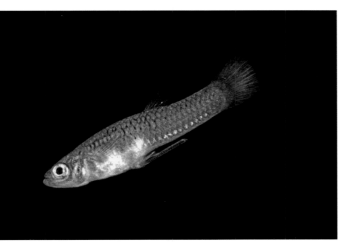

Figure 11.5. A male Amistad Gambusia in an aquarium at the University of Texas. Photograph by Ken Thompson, Fishes of Texas (www.fishesoftexas.org).

PASSENGER PIGEON (*ECTOPISTES MIGRATORIUS*).

Once the most abundant bird in North America, the Passenger Pigeon went extinct in 1914 when the last individual, a female named "Martha," died in captivity at the Cincinnati Zoo (see fig. 3.2). The estimated 3 to 5 billion birds of this species occupied forested habitats east of the Rocky Mountains and usually moved, roosted, and nested in huge flocks. In 1906, ornithologist Alexander Wilson described a sky blackened by a migrating flock that was over a mile wide. Based on the hours required for its passage, he estimated that this aggregation contained 2.2 billion birds and was over 240 miles long. John James Audubon rode through a Passenger Pigeon nesting area that was 3 miles wide and 40 miles in extent. Citizens of Houston mistook a large approaching flock in 1873 for an approaching storm. Passenger Pigeons were regarded as cheap food and were harvested by the tens of thousands during the closing years of the nineteenth century. They were so abundant that a single double-barreled shotgun blast could often down up to 60 birds. Many were shipped by rail to restaurants in large eastern cities. As the birds began to decline from unrestricted hunting, large tracts of forest—the nesting habitats of the species—were being cleared for agriculture. Finally, the populations dipped too low to survive. The last wild Passenger Pigeons in Texas were shot near Galveston in 1900.

IVORY-BILLED WOODPECKER (*CAMPEPHILUS PRINCIPALIS***).** The Ivory-billed Woodpecker, the largest woodpecker in the United States, formerly occupied a range extending from the forests of the eastern third of Oklahoma and Texas to the Atlantic Coast. It was a denizen of old-growth forests, especially those associated with swamps and bottomlands bordering watercourses. These stands contained numerous dying and dead trees that harbored the larvae of wood-boring beetles, the primary food of the woodpecker. Although Ivory-bills supplemented their diet of beetle larvae with fruit, seeds, and insects, each pair required approximately 10 square miles of forested habitat to provide enough food for themselves and their young. When the forests of East Texas were harvested, the woodpecker populations disappeared. The last two Ivory-bill specimens from Texas were taken in Liberty County in 1904, but sightings continued for many years thereafter.

Portions of the Big Thicket were established as a national preserve in 1974, including tracts where the species might have been observed in 1966 and 1968. Undocumented reports of Ivory-bills living in remote areas of Arkansas, Louisiana, Mississippi, and Florida surface periodically and generate both excitement and debate. The species is almost certainly gone, but hope—though fading—prevents reclassification of the species from endangered to extinct.

CAROLINA PARAKEET (*CONUROPSIS CAROLINENSIS***).** The Carolina Parakeet was one of only two parrot species native to the United States. The grackle-sized birds had a bright yellow head with orange cheeks, but their body plumage was grass green (fig. 11.7). Flocks of the parakeet noisily inhabited the deciduous forests and forest edges of eastern North America, and their range extended along river bottoms as far west as Nebraska. Called "Louisiana Parakeets" in Texas, the birds nested in tree cavities and fed on the fruits and seeds of trees and weedy species during the breeding season. Later in the

Figure 11.7. Preserved specimens of the Carolina Parakeet, such as this one at the Field Museum of Natural History in Chicago, are all that remain of this colorful bird. Photograph by James St. John, Creative Commons.

year, the parakeets invaded fruit orchards and fields of cereal crops, where they became serious agricultural pests. Their destructive habits made them targets of persecution, but they were also shot for their colorful feathers, which were used in the millinery trade. Clearing of forests contributed further to their demise. Although they were reported to be numerous along streams in East Texas in 1850, the last parakeet in Texas was killed in Bowie County in 1897. The last Carolina Parakeet died in the Cincinnati Zoo in 1918, and the last Passenger Pigeon died in the same zoo in 1914.

WEST INDIAN MONK SEAL (*NEOMONACHUS TROPICALIS*).

The West Indian Monk Seal was officially declared extinct in 2008 by the National Marine Fisheries Service (fig. 11.8). The species was last seen in 1952 when one was observed in the Caribbean Sea between Jamaica and Nicaragua. The seal may never have been numerous and probably succumbed when overfishing reduced its food supply. The species was also taken for oil. The West Indian Monk Seal rarely entered Texas waters, but it was occasionally seen lounging on breakwater rocks at Bolivar Pass and swimming in the Lower Laguna Madre.

EXTIRPATED SPECIES.

The fauna of Texas also included many species that were extirpated from the state but continue to exist elsewhere. The range of the Black-footed Ferret (*Mustela nigripes*), for example, coincided with that of the Black-tailed Prairie Dog (*Cynomys ludovicianus*), its primary food source. The ferret was driven to extinction by the demise of prairie dog colonies and an outbreak of Sylvatic Plague. The last Black-footed Ferret in Texas was seen in 1953 occupying a prairie dog town in Bailey County. The species was declared extinct in 1979, but the classification changed to endangered in 1981 when a pet dog brought a freshly killed specimen home in Meeteetse, Wyoming. All remaining Black-footed Ferrets in the nearby remnant population were trapped to stock a captive-breeding program. Ferrets produced by that program were reintroduced into prairie dog towns in eight states and Mexico. Over 1,000 ferrets now exist in 18 populations. There are currently no plans to reintroduce Black-footed Ferrets to Texas.

Figure 11.8. The West Indian Monk Seal. Drawing is from a book published in 1887 by George Brown Goode and is in the public domain.

APPENDIX: SCIENTIFIC NAMES OF SPECIES MENTIONED IN THE TEXT

Scientific names of organisms mentioned in the chapters are listed alphabetically by the common name used in the text. Alternate common names and scientific names no longer in use, but occurring in older sources, are also included in brackets; some accounts include informative notations within brackets. Species described in accounts or lists that include both a common name and the corresponding scientific name are not included in this list.

BACTERIA, ALGAE, FUNGI, AND LICHENS

Hantavirus, *Hantavirus* spp.

Lyme Disease, *Borrelia* spp. [Two bacterial species, *B. burgdorferi* and *B. mayonii*, are the infectious agents in North America.]

Muskgrass, *Chara* spp.

Red Tide, *Karenia brevis*

Sylvatic Plague, *Yersinia pestis*

West Nile Fever, *Flavivirus* spp.

PLANTS

Acacia, *Acacia* spp.

Ashe Juniper, *Juniperus ashei*

Aspen, *Populus tremuloides*

Baldcypress, *Taxodium distichum* [Bald Cypress]

Bastard Cabbage, see Turnip Weed

Big Bluestem, *Andropogon gerardii*

Birch, *Betula* spp.

Blackberry, *Rubus* spp.

Blackjack Oak, *Quercus marilandica*

Bluebonnet, *Lupinus* spp.

Bluestem, *Andropogon* and *Schizachyrium* spp.

Brownseed Paspalum, *Paspalum plicatulum*

Buffalograss, *Buchloe dactyloides*

Bulrush, *Scirpus* spp.

Carolina Wolfberry, *Lycium carolinianum*

Cattail, *Typha* spp.

Chinese Tallow, *Triadica sebifera* [*Sapium sebiferum*; exotic]

Coontail, *Ceratophyllum* spp.

Creosotebush, *Larrea tridentata* [Creosote Bush]

Crowberry, *Empetrum nigrum*

Dewberry, *Rubus* spp.

Eastern Cottonwood, *Populus deltoides*

Giant Reed, *Arundo donax*

Glasswort, *Salicornia* spp.

Grama, *Bouteloua* spp.

Havard's Century Plant, *Agave havardiana*

Hickory, *Carya* spp.

Honey Mesquite, *Prosopis glandulosa*

Hydrilla, *Hydrilla verticillata*

Indian Blanket, *Gaillardia pulchella* [Firewheel]

Indiangrass, *Sorghastrum nutans* [Yellow Indiangrass]

Johnsongrass, *Sorghum halepense*

Juniper, *Juniperus* spp.

King Ranch Bluestem, *Bothriochloa ischaemum* var. *songarica*

Lechuguilla, *Agave lechuguilla*

Little Bluestem, *Schizachyrium scoparium* var. *scoparium* [*Andropogon scoparium*]

Live Oak, *Quercus virginiana*

Loblolly Pine, *Pinus taeda*

Longleaf Pine, *Pinus palustris* [Yellow Pine]

Mesquite, *Prosopis* spp.

Mistletoe, *Phoradendron leucarpum*

Morning Glory Tree, *Ipomoea carnea*

Oak, *Quercus* spp.

Ocotillo, *Fouquieria splendens*

Olney's Bulrush, *Schoenoplectus americanus*

Pecan, *Carya illinoinensis*

Pinyon Pine, *Pinus edulis*

Plateau Live Oak, *Quercus fusiformis*

Ponderosa Pine, *Pinus ponderosa*

Post Oak, *Quercus stellata*

Prickly pear, *Opuntia* spp.

River Cane, *Arundinaria gigantea*

Sabal Palm, *Sabal palmetto*

Saltcedar, *Tamarix ramosissima*

Saltwort, *Batis maritima*

Sand Dropseed, *Sporobolus cryptandrus*

Sand Sage, *Artemisia filifolia* [Sand Sagebrush]

Sea Ox-eye, *Borrichia frutescens*

Sedge, *Carex* spp., *Cyperus* spp., and *Eleocharis* spp.

Shinnery Oak, *Quercus havardii* [Havard Shin Oak]

Shortleaf Pine, *Pinus echinata*

Silver Bluestem, *Bothriochloa laguroides* [*Andropogon saccharoides*]

Slash Pine, *Pinus elliottii*

Smooth Cordgrass, *Spartina alterniflora*

Sugar Hackberry, *Celtis laevigata* [Granjeno, Sugarberry, Hackberry]

Tall Fescue, *Festuca arundinacea*

Texas Red Oak, *Quercus buckleyi*

Threeawn, *Aristida* spp.

Turnip Weed, *Rapistrum rugosum*

Water Hyacinth, *Eichhornia crassipes*

Water Tupelo, *Nyssa aquatica*

Wild Grape, *Vitis* spp.

Yucca, *Yucca* spp.

INVERTEBRATE ANIMALS

American Burying Beetle, *Nicrophorus americanus* [Sexton Beetle, Giant Carrion Beetle; *N. grandis, N. virginicis, Silpha orientalis*]

Asian Clam, *Corbicula fluminea*

Blue Crab, *Callinectes sapidus*

Bone Cave Harvestman, *Texella reyesi*

Bracken Bat Cave Meshweaver, *Cicurina venii* [*Cicurella venii*]

Broken Rays Mussel, *Lampsilis reeveiana*

Cave cricket, *Ceuthophilus* spp. [*C. secretus* is common in Central Texas.]

Coffin Cave Mold Beetle, *Batrisodes texanus* [Inner Space Cavern Mold Beetle; *Excavodes texanus*]

Cokendolpher Cave Harvestman, *Texella cokendolpheri* [Robber Baron Cave Harvestman]

Comal Springs Dryopid Beetle, *Stygoparnus comalensis*

Comal Springs Riffle Beetle, *Heterelmis comalensis*

Common Daddy Longlegs, *Leiobunum townsendi*

Coquina (clam), *Donax* spp.

Crayfish, *Cambarus* spp.

Diamond Tryonia, *Pseudotryonia texana* [*Paupertryonia adamantina, Tryonia adamantina*]

Diminutive Amphipod, *Gammarus hyalelloides* [*G. hyalelliodes*]

Dusky Arion, *Arion subfuscus*

Estelline Crab, *Hemigrapsus estellinensis*

European Honey Bee, *Apis mellifera* [exotic; all honey bees in North America were introduced.]

Golden Orb, *Quadrula aurea* [*Unio aureus, Rotundaria aurea, Margaron aurea, Amphinaias aurea*]

Gonzales Tryonia, *Tryonia circumstriata* [*Calipyrgula circimstriata, Tryonia stocktonensis*]

Government Canyon Bat Cave Meshweaver, *Cicurina vespera* [Vesper Cave Spider; *Cicurella vespera*]

Government Canyon Bat Cave Spider, *Tayshaneta microps* [Government Canyon Cave Spider; *Leptoneta microps, Neoleptoneta microps*]

Ground Beetle (no common name), *Rhadine exilis* [*Agonum exile*]

Ground Beetle (no common name), *Rhadine infernalis* [*Agonum infernalis*]

Helotes Mold Beetle, *Batrisodes venyivi* [*Excavodes venyivi*]

Horseshoe Crab, *Limulus polyphemus*

Kretschmarr Cave Mold Beetle, *Texamaurops reddelli*

Macomas, *Macoma* spp.

Madla Cave Meshweaver, *Cicurina madla* [Madla's Cave Spider; *Cicurella madla*]

Ouachita Rock Pocketbook, *Arkansia wheeleri* [Arkansas Rock-pocketbook, Wheeler's Rock-pocketbook, Wheeler's Pearly Mussel; *Arcidens wheeleri*]

Peck's Cave Amphipod, *Stygobromus pecki* [*Stygonectes pecki*]

Pecos Amphipod, *Gammarus pecos*

Pecos Assiminea, *Assiminea pecos*

Phantom Springsnail, *Pyrgulopsis texana* [Reeves County Snail, Phantom Cave Snail; *Cochliopa texana*]

Phantom Tryonia, *Tryonia cheatumi* [Cheatum's Snail, Phantom Springsnail; *Potamopyrgus cheatumi, Lyodes cheatumi*]

Pimpleback, *Quadrula pustulosa*

Pine beetle, *Dendroctonus* spp.

Quilted Melania, *Tarebia granifera* [exotic]

Reddell Cave Harvestman, *Texella reddelli* [Bee Creek Cave Harvestman]

Red Imported Fire Ant, *Solenopsis invicta* [exotic]

Red-rimmed Melania, *Melanoides tuberculata* [exotic]

Robber Baron Cave Meshweaver, *Cicurina baronia* [Robber Baron Cave Spider; *Cicurella baronia*]

Rocky Mountain Grasshopper, *Melanoplus spretus*

Smooth Pimpleback, *Quadrula houstonensis* [Houston Pimpleback, Southern Pimpleback; *Unio houstonensis, Amphinaias houstonensis*]

Surf clams, *Mulina* spp.

Tellin, *Tellina* spp.

Texas Fatmucket, *Lampsilis bracteata* [Fatmucket; *Unio bracteatus, U. rowellii*]

Texas Fawnsfoot, *Truncilla macrodon* [*Unio macrodon*]

Texas Hornshell, *Popenaias popeii* [Hornshell; *Unio popeii, U. veraepacis, Elliptio popeii*]

Texas Pimpleback, *Quadrula petrina* [Pimpleback; *Unio petrinus, U. bolli, Amphinaias petrina*]

Threeridge, *Amblema plicata*

Tooth Cave Ground Beetle, *Rhadine persephone*

Tooth Cave Pseudoscorpion, *Tartarocreagris texana* [Tooth Cave False Scorpion; *Microcreagris texana; Australinocreagris texana*]

Tooth Cave Spider, *Tayshaneta myopica* [*Leptoneta myopica, Neoleptoneta myopica*]

Yellow Shore Crab, *Hemigrapsus oregonensis*
Zebra Mussel, *Dreissena polymorpha*

FISHES

Amistad Gambusia, *Gambusia amistadensis*
Arkansas River Shiner, *Notropis girardi*
Armored catfish, *Hypostomus* spp.
Atlantic Tarpon, *Megalops atlanticus*
Big Bend Gambusia, *Gambusia gaigei*
Bluegill, *Lepomis macrochirus*
Bonnethead Shark, *Sphyrna tiburo*
Caribbean Electric Ray, *Narcine bancroftii* [Bancroft's
    Numbfish, Small Electric Ray, Electric Ray, Spotted
    Torpedo Ray, Torpedofish, Brazilian Electric Ray,
    Trembler; *N. brachypleura, Torpedo brasiliensis*]
Channel Catfish, *Ictalurus punctatus*
Clear Creek Gambusia, *Gambusia heterochir*
Colorado Pikeminnow, *Ptychocheilus lucius*
Comanche Springs Pupfish, *Cyprinodon elegans*
Conchos Pupfish, *Cyprinodon eximius*
Cypress Darter, *Etheostoma proeliare*
Devils River Minnow, *Dionda diaboli*
Duskystripe Shiner, *Luxilus pilsbryi*
Fountain Darter, *Etheostoma fonticola* [*Alvarius fonticola,
    Microperca fonticola*]
Freshwater Drum, *Aplodinotus grunniens*
Golden Shiner, *Notemigonus crysoleucas*
Gray Redhorse, *Moxostoma congestum*
Great Hammerhead, *Sphyrna mokarran* [Squat-headed
    Hammerhead Shark; *Zygaena mokarran*]
Green Sunfish, *Lepomis cyanellus*
Greenthroat Darter, *Etheostoma lepidum*
Guadalupe Bass, *Micropterus treculii*
Guadalupe Darter, *Percina apristis*
Largemouth Bass, *Micropterus salmoides*
Largespring Gambusia, *Gambusia geiseri*
Largetooth Sawfish, *Pristis pristis* [Leichhardt's Sawfish,
    Freshwater Sawfish; *P. microdon, P. perotteti, P.
    antiquorum, P. zephyreus*]
Leon Springs Pupfish, *Cyprinodon bovinus*
Lesser Electric Ray, *Narcine bancroftii*
Mosquitofish, *Gambusia affinis*
Orangethroat Darter, *Etheostoma spectabile*
Pecos Gambusia, *Gambusia nobilis*
Pecos Pupfish, *Cyprinodon pecosensis*
Red River Pupfish, *Cyprinodon rubrifluviatilis*
Red Shiner, *Cyprinella lutrensis*
Rio Grande Silvery Minnow, *Hybognathus amarus*
River Carpsucker, *Carpiodes carpio*
Sandlance, *Ammodytes* spp.
San Marcos Gambusia, *Gambusia georgei*

Scalloped Bonnethead, *Sphyrna corona*
Scalloped Hammerhead, *Sphyrna lewini* [Bronze
    Hammerhead, Kidney-headed Shark, Southern
    Hammerhead Shark; *Zygaena lewini*]
Sharpnose Shiner, *Notropis oxyrhynchus*
Sheepshead Minnow, *Cyprinodon variegatus*
Smalleye Shiner, *Notropis buccula* [*Notropis bairdi
    buccula*]
Smalltooth Sawfish, *Pristis pectinata* [Wide Sawfish]
Smooth Hammerhead, *Sphyrna zygaena*
Texas Logperch, *Percina carbonaria*
Western Mosquitofish, *Gambusia affinis*

AMPHIBIANS AND REPTILES

American Alligator, *Alligator mississippiensis*
Austin Blind Salamander, *Eurycea waterlooensis*
Barton Springs Salamander, *Eurycea sosorum*
Brown Anole, *Anolis sagrei* [Bahamian Anole; exotic]
Bullsnake, *Pituophis catenifer sayi*
Georgetown Salamander, *Eurycea naufragia*
Green Anole, *Anolis carolinensis*
Green Sea Turtle, *Chelonia mydas* [Spanish: *Tortuga
    Blanca, Tortuga Verde*]
Hawksbill Sea Turtle, *Eretmochelys imbricata* [Tortoise-
    shell Turtle, Caret; Spanish: *Tortuga Carey*]
Houston Toad, *Anaxyrus houstonensis* [*Bufo houstonensis*]
Jollyville Plateau Salamander, *Eurycea tonkawae*
Kemp's Ridley Sea Turtle, *Lepidochelys kempii* [Atlantic
    Ridley, Mexican Ridley, Gulf Ridley; Spanish: *Tortuga
    Lora*]
Leatherback Sea Turtle, *Dermochelys coriacea* [Leathery
    Turtle, Luth, Trunkback Turtle, Trunk Turtle, Coffin-
    back; Spanish: *Tortuga Laúd*]
Loggerhead Sea Turtle, *Caretta caretta* [Spanish: *Tortuga
    Caguama*]
Louisiana Pinesnake, *Pituophis ruthveni* [Louisiana
    Pine Snake, Pine Snake, Pinesnake; *P. melanoleucus
    ruthveni*]
Massasauga, *Sistrurus catenatus*
Salado Salamander, *Eurycea chisholmensis*
San Marcos Salamander, *Eurycea nana* [Spring Lake
    Salamander]
Texas Blind Salamander, *Eurycea rathbuni*
Texas Indigo Snake, *Drymarchon melanurus erebennus*
    [*Drymarchon corais erebennus*]
Texas Tortoise, *Gopherus berlandieri*
Western Massasauga, *Sistrurus catenatus tergeminus*

BIRDS

American Golden-Plover, *Pluvialis dominica*
American Pipit, *Anthus rubescens*

Attwater's Prairie-Chicken, *Tympanuchus cupido attwateri* [Louisiana Prairie Hen; *T. attwateri, T. americanus attwateri*]

Bald Eagle, *Haliaeetus leucocephalus*

Barred Owl, *Strix varia*

Black-and-White Warbler, *Mniotilta varia*

Black-capped Vireo, *Vireo atricapilla* [*V. atricapillus*]

Blue-headed Vireo, *Vireo solitarius*

Blue-winged Teal, *Anas discors*

Brown-headed Cowbird, *Molothrus ater*

Brown Pelican, *Pelecanus occidentalis*

California Condor, *Gymnogyps californicus*

California Least Tern, *Sternula antillarum browni* [*Sterna a. browni*]

California Spotted Owl, *Strix occidentalis occidentalis*

Canary, *Serinus canaria domestica* [a domesticated species used by coal miners]

Carolina Parakeet, *Conuropsis carolinensis* [extinct]

Cave Swallow, *Petrochelidon fulva* [*Hirundo fulva*]

Coastal Least Tern, see Eastern Least Tern

Common Raven, *Corvus corax*

Crissal Thrasher, *Toxostoma crissale*

Eastern Bluebird, *Sialia sialis*

Eastern Least Tern, *Sternula antillarum antillarum* [Coastal Least Tern; *Sterna a. antillarum*]

Eskimo Curlew, *Numenius borealis* [Doughbird, Doe-bird, Prairie Pigeon, Fute, Little Curlew; *Mesoscolopax borealis*; likely extinct]

European Starling, *Sturnus vulgaris* [exotic]

Gadwall, *Mareca strepera*

Gambel's Quail, *Callipepla gambelii*

Golden-cheeked Warbler, *Setophaga chrysoparia* [*Dendroica chrysoparia*; Spanish: *Chipe Mejilla Dorada*; French: *Paruline à Dos Noir*]

Golden Eagle, *Aquila chrysaetos*

Great Auk, *Pinguinus impennis* [extinct]

Great Egret, *Ardea alba* [*Casmerodius albus*]

Greater Prairie-Chicken, *Tympanuchus cupido cupido*

Greater Roadrunner, *Geococcyx californicus*

Great Horned Owl, *Bubo virginianus*

Great White Heron, *Ardea herodias occidentalis*

Heath Hen, *Tympanuchus cupido cupido* [extinct subspecies of the Greater Prairie-Chicken]

Hudsonian Curlew, *Numenius phaeopus hudsonicus*

Interior Least Tern, *Sternula antillarum athalossus* [*Sterna antillarum athalossus, Sterna albifrons*; Spanish: *Charrancito Americano*]

Ivory-billed Woodpecker, *Campephilus principalis* [likely extinct]

Labrador Duck, *Camptorhynchus labradorius* [extinct]

Lesser Prairie-Chicken, *Tympanuchus pallidicinctus*

[Lesser Prairie Hen; *Cupidonia cupido pallidicincta, C. c. pallidicinctus*]

Mallard, *Anas platyrhynchos*

Mexican Spotted Owl, *Strix occidentalis lucida* [Spotted Owl; *Syrnium occidentale*; Spanish: *Búho Manchado, Cárabo Californiano*]

Northern Aplomado Falcon, *Falco femoralis septentrionalis* [Orange-chested Hobby; *Falco fusco-coerulescens, F. fuscocaerulescens*; Spanish: *Halcón Aplomado, Halcón Fajado*; Portuguese: *Falcäo-de-Coleira*]

Northern Bobwhite, *Colinus virginianus*

Northern Pintail, *Anas acuta*

Northern Spotted Owl, *Strix occidentalis caurina*

Osprey, *Pandion haliaetus*

Passenger Pigeon, *Ectopistes migratorius* [extinct]

Peregrine Falcon, *Falco peregrinus*

Piping Plover, *Charadrius melodus* [Ringneck, Sand Plover, Clam Bird, Belted Piping Plover, Mourning Bird, Beach Plover; *Aegialitis meloda, A. melodus, Hiaticula meloda, Charadrius okeni*]

Red-cockaded Woodpecker, *Picoides borealis*

Red-crowned Parrot, *Amazona viridigenalis* [Red-crowned Amazon, Green-cheeked Amazon, Mexican Red-headed Parrot; Spanish: *Loro Tamaulipeco, Loro Cabeza Roja*]

Red Knot, *Calidris canutus*

Sagebrush Sparrow, *Artemisiospiza nevadensis* [*Amphispiza nevadensis*]

Sandhill Crane, *Antigone canadensis* [*Grus canadensis*]

Sandpipers, *Calidris* spp.

Sharp-tailed Grouse, *Tympanuchus phasianellus*

Snowy Plover, *Charadrius nivosus*

Southwestern Willow Flycatcher, *Empidonax traillii extimus* [Little Flycatcher, Little Western Flycatcher, Traill's Flycatcher; *Empidonax alnorum*]

Sprague's Pipit, *Anthus spragueii* [Missouri Skylark, Prairie Skylark, Titlark; *Anthus spraguei, Alauda spragueii*; Spanish: *Bisbita Llanera*]

Western Atlantic Red Knot, *Calidris canutus rufa* [Knot, Red Sandpiper, Red-breasted Sandpiper, Red-breasted Plover, Beach Robin, Canute's Sandpiper, Silver Plover; *Calidris islandica, Canutus canutus rufa, Charadrius utopiensis, Tringa canutus*]

Western Meadowlark, *Sturnella neglecta*

Western Yellow-billed Cuckoo, *Coccyzus americanus occidentalis* [Rain Crow, Storm Crow]

Whimbrel, *Numenius phaeopus phaeopus*

Whooping Crane, *Grus americana* [White Crane; Spanish: *Grulla Gritona, Grulla Americana, Grulla Trompetera*]

Willow Flycatcher, *Empidonax traillii*

Wilson's Warbler, *Wilsonia pusilla*
Wood Stork, *Mycteria americana*
Zone-tailed Hawk, *Buteo albonotatus*

MAMMALS
African Elephant, *Loxodonta africana* or *L. cyclotis*
African Lion, *Panthera leo* [*Felis leo*]
American Beaver, *Castor canadensis*
American Bison, *Bos bison* [incorrectly known as "Buffalo"]
American Black Bear, *Ursus americanus*
Baird's Pocket Gopher, *Geomys breviceps*
Black Bear, *Ursus americanus*
Black-footed Ferret, *Mustela nigripes*
Black-tailed Jackrabbit, *Lepus californicus*
Black-tailed Prairie Dog, *Cynomys ludovicianus*
Blue Whale, *Balaenoptera musculus*
Bryde's Whale, *Balaenoptera edeni*
Caribbean Monk Seal, *Monachus tropicalis* [West Indian Monk Seal; extinct]
Cheetah, *Acinonyx jubatus*
Collared Peccary, *Pecari tajacu* [Javelina]
Common Muskrat, *Ondatra zibethicus*
Coyote, *Canis latrans*
Crabeater Seal, *Lobodon carcinophaga*
Desert Bighorn Sheep, *Ovis canadensis nelsoni*
Dugong, *Dugong dogon*
Eastern Cottontail, *Sylvilagus floridanus*
Elk, *Cervus canadensis* [Wapiti; *C. elaphus*]
Feral Pig, *Sus scrofa* [Feral Hog, Wild Boar]
Fin Whale, *Balaenoptera physalus* [Finback Whale, Finner, Razorback, Common Rorqual; Spanish: *Rorcual Común*]
German Shepherd, *Canis familiaris* [a domestic dog]
Gray Fox, *Urocyon cinereoargenteus*
Gray Wolf, *Canis lupus*
Gulf Coast Jaguarundi, *Puma yagouaroundi cacomotli* [Otter Cat, Weasel Cat, Eyra Cat; *Felis yagouaroundi*, *Herpailurus yagouaroundi*; Spanish: *Yaguarundi, Tigrillo, Onza, Gato Eyra*]
Harp Seal, *Pagophilus groenlandicus* [Saddleback Seal]
Hispid Cotton Rat, *Sigmodon hispidus*
Humpback Whale, *Megaptera novaeangliae* [Spanish: *Rorcual Jorobado*]
Jaguar, *Panthera onca* [*Felis onca*; Spanish: *Tigre Real, Tigre Americano, Otorongo, Yaguar, Yaguarete*]

Jaguarundi, *Puma yagouaroundi* [*Herpailurus yagouaroundi*]
Lesser Long-nosed Bat, *Leptonycteris yerbabuenae*
Louisiana Black Bear, *Ursus americanus luteolus* [*Euarctos americanus luteolus*]
Manatee, see West Indian Manatee
Mexican Free-tailed Bat, *Tadarida brasiliensis*
Mexican Gray Wolf, *Canis lupus baileyi* [Lobo, Mexican Wolf, Timber Wolf]
Mexican Long-nosed Bat, *Leptonycteris nivialis* [Big Long-nosed Bat, Greater Long-nosed Bat, Nectar Bat; Spanish: *Murciélago Hocicudo Mayor*]
Mexican Long-tongued Bat, *Choeronycteris mexicana*
Minke Whale, *Balaenoptera bonaerensis*
Mountain Lion, *Puma concolor* [Cougar, Puma, Panther; *Felis concolor*]
Mule Deer, *Odocoileus hemionus*
Nutria, *Myocastor coypus*
Ocelot, *Leopardus pardalis* [Leopard Cat, Little Leopard, Little Tiger, Dwarf Leopard; *Felis pardalis*; Spanish: *Manigordo, Ocelote*]
Packrat, *Neotoma* spp.
Pocket gopher, *Geomys* spp.
Polar Bear, *Ursus maritimus*
Pronghorn, *Antilocapra americana*
Raccoon, *Procyon lotor*
Red Wolf, *Canis rufus* [Florida Wolf, Mississippi Valley Wolf; *C. lupus rufus, C. niger*]
Reticulated Giraffe, *Giraffa camelopardis reticulata*
Sea Mink, *Neovison macrodon* [extinct]
Sea Otter, *Enhydra lutris*
Sinaloan Jaguarundi, *Herpailurus yagouaroundi tolteca* [*Puma y. tolteca*]
Sperm Whale, *Physeter macrocephalus* [*Physeter catodon, P. australasianus;*
Spanish: *Cachalote*]
Striped Skunk, *Mephitis mephitis*
Texas Gray Wolf, *Canis lupus nubilus* [extinct]
Texas Kangaroo Rat, *Dipodomys elator*
Walrus, *Odobenus rosmarus*
West Indian Manatee, *Trichechus manatus* [Manatee, Caribbean Manatee, Florida Manatee, Sea Cow; *Manatus latirostris*]
West Indian Monk Seal, *Neomonachus tropicalis*
White-tailed Deer, *Odocoileus virginianus*

SELECTED REFERENCES

The most significant references consulted during the preparation of this book, as well as a few suggestions for additional readings of related interest, are presented in a format that mirrors the headings and subheadings in each chapter. References consulted for multiple chapters are listed below and are not repeated elsewhere.

Ammerman, L. K., C. L. Hice, and D. J. Schmidly. 2012. Bats of Texas. Texas A&M University Press, College Station. 305 pp.

Campbell, L. 1995. Endangered and threatened animals of Texas. Texas Parks and Wildlife Department, Endangered Resources Branch, Austin. 129 pp.

Correll, D. S., and M. C. Johnston. 1970. The manual of the vascular plants of Texas. Texas Research Foundation, Renner. 1881 pp.

Graves, J. 2002. Texas rivers. Texas Parks and Wildlife Press, Austin. 144 pp.

Griffith, G., S. Bryce, J. Omernik, and A. Rogers. 2007. Ecoregions of Texas. Texas Commission on Environmental Quality, Austin. 125 pp.

Hoese, H. D., and R. H. Moore. 1998. Fishes of the Gulf of Mexico: Texas, Louisiana, and adjacent waters. 2nd rev. ed. Texas A&M University Press, College Station. 416 pp.

Huser, V. 2000. Rivers of Texas. Texas A&M University Press, College Station. 264 pp.

Jones, J. K., Jr. 1993. The concept of threatened and endangered species as applied to Texas mammals. Texas Journal of Science 45:115-128.

Larkin, T. J., and G. W. Bomar. 1983. Climatic atlas of Texas. Texas Department of Water Resources LP-192:1-151.

Lockwood, M. W., and B. Freeman. 2014. The TOS handbook of Texas birds. 2nd ed. Texas A&M University Press, College Station. 403 pp.

Oberholser, H. C. 1974. The bird life of Texas. Edited by E. G. Kincaid Jr. 2 vols. University of Texas Press, Austin. 1069 pp.

Primrack, R. B. 2014. Essentials of conservation biology. 6th ed. Sinauer Associates, New York. 603 pp.

Schmidly, D. J. 2002. Texas natural history: A century of change. Texas Tech University Press, Lubbock. 534 pp.

Schmidly, D. J., and R. D. Bradley. 2016. The mammals of Texas. 7th ed. University of Texas Press, Austin. 720 pp.

Sellards, E. H., W. S. Adkins, and F. B. Plummer. 1932. The geology of Texas. University of Texas Bulletin 3232 (1): 1-1007.

Spearing, D. 1991. Roadside geology of Texas. Mountain Press, Missoula, MT. 418 pp.

Thomas, C., T. H. Bonner, and B. G. Whiteside. 2007. Freshwater fishes of Texas. Texas A&M University Press, College Station. 202 pp.

Werler, J. E., and J. R. Dixon. 2000. Texas snakes: Identification, distribution, and natural history. University of Texas Press, Austin. 437 pp.

Wilson, E. O., ed. 1988. Biodiversity. National Academy Press, Washington, DC. 521 pp.

Chapter One

Becher, M. A., J. L. Osborne, P. Thorbeck, et al. 2013. Towards a systems approach for understanding honeybee decline: A stocktaking and synthesis of existing models. Journal of Applied Ecology 50:868-880.

Carson, R. 1962. Silent spring. Houghton Mifflin, New York. 368 pp.

Ceballos, G., P. R. Ehrlich, and R. Dirzo. 2017. Biological annihilation via the ongoing sixth mass extinction signaled by vertebrate population losses and declines. Proceedings of the National Academy of Sciences (online). doi:10.1073/pnas.170494114.

Chivan, E., and A. Bernstein, eds. 2008. Sustaining life: How human health depends on biodiversity. 3rd ed. Oxford University Press, New York. 568 pp.

Federico, P., T. G. Hallum, G. F. McCracken, et al. 2008. Brazilian Free-tailed Bats as insect pest regulators in transgenic and conventional cotton crops. Ecological Applications 18:826-837.

Honey, M. 2008. Ecotourism and sustainable development: Who owns paradise? 2nd ed. Island Press, Washington, DC. 568 pp.

Kolbert, E. 2015. The sixth extinction: An unnatural history. Henry Holt, New York. 336 pp.

Li, J., A. D. Celiz, L. Yang, et al. 2017. Tough adhesives for diverse wet surfaces. Science 357 (6349): 378-381.

Luck, G. W., G. C. Daly, and P. R. Ehrlich. 2003. Population diversity and ecosystem services. Trends in Ecology and Evolution 18:331–336.

Maxwell, S. L., R. A. Fuller, T. M. Brooks, and J. E. M. Watson. 2016. Biodiversity: The ravages of guns, nets, and bulldozers. Nature 536:143–145.

Mills, L. S., M. E. Soule, and D. F. Doak. 1993. The keystone species concept in ecology and conservation. BioScience 43 (4): 219–224.

Pearson, R. G. 2016. Reasons to conserve nature. Trends in Ecology and Evolution 31:366–371.

Potts, S., and J. Biesmeijer. 2010. Global pollinator declines: Trends, impacts, and drivers. Trends in Ecology 20:367–373.

Sala, O. E., L. A. Meyerson, and C. Parmesan, eds. 2009. Biodiversity change and human health: From ecosystem services to spread of disease. Island Press, Washington, DC. 320 pp.

Wake, D. B., and V. T. Vredenburg. 2008. Are we in the midst of the sixth mass extinction? A view from the world of amphibians. Proceedings of the National Academy of Sciences 105:11466–11473.

Chapter Two

Bolen, E. G., and W. L. Robinson. 2003. Wildlife ecology and management. 5th ed. Pearson Benjamin Cummings, San Francisco. 634 pp.

Fisher, J., N. Simon, and J. Vincent. 1969. Wildlife in danger. Viking Press, New York. 368 pp.

Kotiaho, J. S., V. Kaitala, A. Komonen, et al. 2005. Predicting the risk of extinction from shared ecological characteristics. Proceedings of the National Academy of Sciences 102:1963–1967.

Chapter Three

Czech, B., and P. R. Krausmann. 2001. The Endangered Species Act: History, conservation, biology, and public policy. Johns Hopkins University Press, Baltimore, MD. 212 pp.

Goble, D. D., J. M. Scott, and F. W. Davis, eds. 2005. The Endangered Species Act at thirty. Vol. 1, Renewing the conservation promise. Island Press, Washington, DC. 392 pp.

———. 2005. The Endangered Species Act at thirty. Vol. 2, Conserving biodiversity in human-dominated landscapes. Island Press, Washington, DC. 376 pp.

Hornaday, W. T. 2002. The extermination of the American Bison. Smithsonian Institution Scholarly Press, Washington, DC. 240 pp. [Reprint of the original 1887 publication.]

Roman, J. 2011. Listed: Dispatches from America's Endangered Species Act. Harvard University Press, Cambridge, MA. 368 pp.

Sadasivam, N. 2017. Endangered science: Is Texas rigging the process to keep species off the federal list? Texas Observer (June).

Stanford Environmental Law Society. 2001. The Endangered Species Act (Stanford Environmental Law Society Handbook). Stanford University Press, Stanford, CA. 312 pp.

Thomas, J. W., and D. E. Toweil. 1982. Elk of North America: Ecology and management. Stackpole Books, Harrisburg, PA. 720 pp.

Trefethen, J. B. 1975. An American crusade for wildlife. Boone and Crockett Club, Winchester Press, New York. 421 pp.

US Fish and Wildlife Service and National Marine Fisheries Service. 1996. Policy regarding the recognition of distinct vertebrate population segments under the Endangered Species Act. Federal Register 61 (26): 4722–4725.

Wake, D. B., and V. T. Vredenburg. 2008. Are we in the midst of the sixth mass extinction? A view from the world of amphibians. Proceedings of the National Academy of Sciences 105:11466–11473.

Chapter Four

Bezanson, D. 2000. Natural vegetation types of Texas and their representation in conservation areas. Unpublished master's thesis, University of Texas, Austin. 215 pp.

Blair, F. 1950. The biotic provinces of Texas. Texas Journal of Science 2:93–117.

Bomar, G. W. 1995. Texas weather. 2nd rev. ed. University of Texas Press, Austin. 287 pp.

Carter, W. T. 1931. The soils of Texas. Texas Agricultural Experiment Station Bulletin 431:1–190.

Chapman, B. R., and E. G. Bolen. 2018. The natural history of Texas. Texas A&M University Press, College Station. 373 pp.

Godfrey, C., G. S. McKee, and H. Oats. 1973. General soils map of Texas. Texas Agricultural Experiment Station Miscellaneous Publication MP-1304.

Gould, F. W. 1969. Texas plants, a checklist and ecological summary. Revised. Texas Agricultural Experiment Station, Texas A&M University, College Station. 121 pp.

Griffith, G., S. Bryce, J. Omernik, and A. Rogers. 2007. Ecoregions of Texas. Texas Commission on Environmental Quality, Austin. 125 pp.

Roe, R. 2017. Birding the corners. Texas Parks and Wildlife Magazine 75 (7): 40–46.

Chapter Five

FRESHWATER MUSSELS

Galbraith, H. S., D. E. Spooner, and C. C. Vaughn. 2008. Status of rare and endangered freshwater mussels in southeastern Oklahoma. Southwestern Naturalist 53:45–50.

Haag, W. R. 2012. North American freshwater mussels: Natural history, ecology, and conservation. Cambridge University Press, New York. 505 pp.

Howells, R. G. 1997. New fish hosts for nine freshwater mussels (Bivalvia: Unionidae) in Texas. Texas Journal of Science 49:255–258.

———. 2013. Field guide to Texas freshwater mussels. BioStudies, Kerrville, TX. 141 pp.

Howells, R. G., R. W. Neck, and H. D. Murray. 1996. Freshwater mussels of Texas. Texas Parks and Wildlife Department, Austin. 218 pp.

Howells, R. G., C. R. Randklev, and M. S. Johnson. 2011. Mantle flap variation in Texas Fatmucket (*Lampsilis bracteata*). Ellipsaria 13:14–16.

Levine, T. D., B. K. Lang. and D. J. Berg. 2012. Physiological and ecological hosts of *Popenaias popeii* (Bivalvia: Unionidae): Laboratory studies identify more hosts than field studies. Freshwater Biology 57:1854–1864.

Randklev, C. R., M. S. Johnson, E. T. Tsakiris, et al. 2012. False Spike, *Quadrula mitchelli* (Bivalvia: Unionidae), is not extinct: First account of a live population in over 30 years. American Malacological Bulletin 30: 327–328.

Randklev, C. R., E. Tsakiris, R. G. Howells, et al. 2013. Distribution of extant populations of *Quadrula mitchelli* (False Spike). Ellipsaria 15:18–21.

Strenth, N. E., R. G. Howells, and A. Correa-Sandoval. 2001. New records of the Texas Hornshell *Popenaias popeii* (Bivalvia: Unionidae) from Texas and northern Mexico. Texas Journal of Science 56:223–230.

Williams, J. D., A. E. Bogan, and J. T. Garner. 2008. Freshwater mussels of Alabama and the Mobile Basin in Georgia, Mississippi, and Tennessee. University of Alabama Press, Tuscaloosa. 908 pp.

Williams, J. D., M. L. Warren Jr., K. S. Cummings, et al. 1993. Conservation status of freshwater mussels of the United States and Canada. Fisheries 18:6–22.

THE FOUNTAIN DARTER

Alexander, M. L., and C. T. Phillips. 2012. Habitats used by the endangered Fountain Darter (*Etheostoma fonticola*) in the San Marcos River, Hays County, Texas. Southwestern Naturalist 57:449–452.

Craig, C. A., K. A. Kollaus, K. P. K. Behen, and T. H. Bonner. 2016. Relationships among spring flow, habitats, and fishes within evolutionary refugia of the Edwards Plateau. Ecosphere 7:1–13.

Dammeyer, N. T., C. T. Phillips, and T. H. Bonner. 2013. Site fidelity and movement of *Etheostoma fonticola* with implications to endangered species management. Transactions of the American Fisheries Society 142:1049–1057.

Kollaus, K. A., K. P. K. Behen, T. C. Heard, et al. 2015. Influence of urbanization on a karst terrain stream and fish community. Urban Ecosystems 18:293–320.

Linam, G. W., K. B. Mayes, and K. S. Saunders. 1993. Habitat utilization and population size estimate of Fountain Darters, *Etheostoma fonticola*, in the Comal River, Texas. Texas Journal of Science 45:341–348.

Nichols, H. 2015. Habitat mediated effects on *Etheostoma fonticola* reproduction. Master's thesis, Texas State University, San Marcos. 44 pp.

Olsen, J. B., A. P. Kinziger, J. K. Wenburg, et al. 2016. Genetic diversity and divergence in the Fountain Darter (*Etheostoma fonticola*): Implications for conservation of an endangered species. Conservation Genetics 17:1393–1404.

Scanes, C. M. 2016. Fish community and habitat assessments within an urbanized spring complex of the Edwards Plateau. Master's thesis, Texas State University, San Marcos.

Schenck, J., and B. Whiteside. 1977. Reproduction, fecundity, sexual dimorphism and sex ratio of *Etheostoma fonticola* (Osteichthyes: Percidae). American Midland Naturalist 98:365–375.

TEXAS MOSQUITOFISHES

Bednarz, J. C. 1979. Ecology and status of the Pecos Gambusia, *Gambusia nobilis* (Poeciliidae), in New Mexico. Southwestern Naturalist 24 (2): 311–322.

Edwards, R. J. 1999. Ecological profiles for selected stream-dwelling Texas freshwater fishes II. Report to the Texas Water Development Board. 69 pp.

———. 2001. Ecological profiles for selected stream-dwelling Texas freshwater fishes III. Report to the Texas Water Development Board. 59 pp.

Hubbs, C. L. 1929. Studies of the fishes of the order Cyprinodontes. VIII. *Gambusia gaigei*, a new species from the Rio Grande. Occasional Papers of the Museum of Zoology, University of Michigan 198:1–11.

Hubbs, C., and H. J. Brodrick. 1963. Current abundance of *Gambusia gaigei*, an endangered fish species. Southwestern Naturalist 8 (1): 46–48.

Hubbs, C., R. J. Edwards, and G. P. Garrett. 2002. Threatened fishes of the world: *Gambusia nobilis* Baird and Girard, 1853 (Poeciliidae). Environmental Biology of the Fishes 64:428.

Hubbs, C., and A. E. Peden. 1969. *Gambusia georgei* sp. nov. from San Marcos, Texas. Copeia 1969:357–364.

Hubbs, C., and V. G. Springer. 1957. A revision of the *Gambusia nobilis* species group, with descriptions of three new species, and notes on their variation, ecology and evolution. Texas Journal of Science 9:279–327.

Rohde, F. C. 1980. *Gambusia heterochir* (Hubbs), Clear Creek Gambusia. P. 543 *in* Atlas of North American freshwater fishes (D. S. Lee et al., eds.). North Carolina State Museum of Natural History, Raleigh. 854 pp.

Warren, M. L., Jr., B. M. Burr, S. J. Walsh, et al. 2000. Diversity, distribution, and conservation status of the native freshwater fishes of the southern United States. Fisheries 25:7–29.

RIVER AND STREAM MINNOWS

Àlo, D., and T. E. Turner. 2005. Effects of habitat fragmentation on effective population size in the endangered Rio Grande Silvery Minnow. Conservation Biology 19:1138–1148.

Bestgen, K. R., and S. P. Platania. 1991. Status and conservation of the Rio Grande Silvery Minnow, *Hybognathus amarus*. Southwestern Naturalist 36:225–232.

Bonner, T. H., and G. R. Wilde. 2000. Changes in the Canadian River fish assemblage associated with reservoir construction. Journal of Freshwater Ecology 15:189–198.

Cantu, N. E., and K. O. Winemiller. 1997. Structure and habitat associations of Devils River fish assemblages. Southwestern Naturalist 42:265–278.

Garrett, G. P., R. J. Edwards, and C. Hubbs. 2004. Discovery of a new population of Devils River Minnow (*Dionda diaboli*), with implications for conservation of the species. Southwestern Naturalist 49:435–441.

Garrett, G. P., R. J. Edwards, and A. H. Price. 1992. Distribution and status of the Devils River Minnow, *Dionda diaboli*. Southwestern Naturalist 37:259–267.

Harrell, H. L. 1978. Response of the Devils River (Texas) fish community to flooding. Copeia 1978:60–68.

Hubbs, C. 1951. Observations on the breeding of *Dionda episcopa serena* in the Nueces River, Texas. Texas Journal of Science 3:490–492.

Hubbs, C., and G. P. Garrett. 1990. Reestablishment of *Cyprinodon eximius* (Cyprinodontidae) and status of *Dionda diaboli* (Cyprinidae) in the vicinity of Dolan Creek, Val Verde Co., Texas. Southwestern Naturalist 35:446–478.

Lopez-Fernandez, H., and K. O. Winemiller. 2005. Status of *Dionda diaboli* and report of established populations of exotic fish species in lower San Felipe Creek, Val Verde County, Texas. Southwestern Naturalist 50:246–251.

Phillips, C. T., R. Gibson, and J. N. Fries. 2011. Spawning behavior and nest association by *Dionda diaboli* in the Devils River, Texas. Southwestern Naturalist 56:108–112.

Robinson, D. T. 1959. The ichthyofauna of the lower Rio Grande, Texas and Mexico. Copeia 1959:253–256.

Sublette, J. E., M. D. Hatch, and M. Sublette. 1990. The fishes of New Mexico. University of New Mexico Press, Albuquerque. 393 pp.

PRAIRIE STREAM SHINERS

Bestgen, K. R., S. P. Platania, J. E. Brooks, and D. L. Propst. 1989. Dispersal and life history traits of *Notropis girardi* (Cypriniformes: Cyprinidae), introduced into the Pecos River, New Mexico. American Midland Naturalist 122:228–235.

Bielawsk, J. P., and J. R. Gold. 2001. Phylogenetic relationships of cyprinid fishes in subgenus *Notropis* inferred from nucleotide sequences of the mitochondrially encoded cytochrome b gene. Copeia 2001:656–667.

Bonner, T. H., and G. R. Wilde. 2000. Changes in the Canadian River fish assemblage associated with reservoir construction. Journal of Freshwater Ecology 15:189–198.

Durham, B. W., and G. R. Wilde. 2009. Effects of streamflow and intermittency on the reproductive success of two broadcast-spawning cyprinid fishes. Copeia 2009:21–28.

Ostrand, K. G., and G. R. Wilde. 2001. Temperature, dissolved oxygen, and salinity tolerances of five prairie stream fishes and their role in explaining fish assemblage patterns. Transactions of the American Fisheries Society 130:742–749.

Polivka, K. M. 1999. The microhabitat distribution of the Arkansas River Shiner, *Notropis girardi*: A habitat-mosaic approach. Environmental Biology of Fishes 55: 265–278.

Wilde, G. R., and A. C. Urbanczyk. 2013. Relationship between river fragment length and persistence of two imperiled Great Plains cyprinids. Journal of Freshwater Ecology 28:445–451.

Chapter Six

Barr, T. C. 1967. Observations on the ecology of caves. American Naturalist 101:475–491.

Culver, D. C., and T. Pipan. 2009. The biology of caves and other subterranean habitats. Oxford University Press, New York. 256 pp.

Elliott, W. R. 2000. Conservation of the North American cave and karst biota. Pp. 665–689 in Subterranean ecosystems. Ecosystems of the World, 30 (H. Wilkens, D. C. Culver, and W. F. Humphreys, eds.). Elsevier, Amsterdam, the Netherlands. 808 pp.

Holsinger, J. R. 1988. Troglobites: The evolution of cave-dwelling organisms. American Scientist 76:147–153.

Owen, J. D., S. van Kampen-Lewis, K. White, and C. Crawford. 2016. Preliminary study of Central Texas cave cricket monitoring (Orthoptera: Rhaphidophoridae). Southwestern Naturalist 61:265–269.

Romero, A. 2009. Cave biology: Life in darkness. Cambridge University Press, New York. 306 pp.

Taylor, S. J., J. K. Krejca, and M. L. Denight. 2005. Foraging range and habitat use of Ceuthophilus secretus (Orthoptera: Rhaphidophoridae), a key trogloxene in Central Texas cave communities. American Midland Naturalist 154:97–114.

Taylor, S. J., J. D. Wecjstein, D. M. Takiya, et al. 2007. Phylogeography of cave crickets (Ceuthophilus spp.) in Central Texas: A keystone taxon for the conservation and management of federally listed endangered cave arthropods. Illinois Natural History Survey Technical Report 2007 (58): 1–45.

Veni, G. 1994. Hydrogeology and evolution of caves and karst in the southwestern Edwards Plateau, Texas. Pp. 13–30 in The caves and karst of Texas (W. R. Elliott and G. Veni, eds.). National Speleological Society, Huntsville, AL. 252 pp.

TOOTH CAVE PSEUDOSCORPION

Mitchell, R. W., and J. R. Reddell. 1971. The invertebrate fauna of Texas caves. Pp. 35–40 in Natural history of Texas caves (E. L. Lundelius Jr. and B. H. Slaughter, eds.). Gulf Natural History Publishing, Dallas, TX. 174 pp.

Muchmore, W. B. 1992. Cavernicolous pseudoscorpions from Texas and New Mexico (Arachnida: Pseudoscorpionida). Texas Memorial Museum, Speleological Monographs 3:127–153.

Reddell, J. R. 1994, The cave fauna of Texas with special reference to the western Edwards Plateau. Pp. 31–50 in The caves and karst of Texas (W. R. Elliott and G. Veni, eds.) National Speleological Society, Huntsville, AL. 252 pp.

Rowland, J. M., and J. R. Reddell. 1976. Annotated checklist of the arachnid fauna of Texas (excluding Acarida and Araneida). Occasional Papers of the Museum, Texas Tech University 38:1–25.

US Fish and Wildlife Service. 2009. Tooth Cave Spider (Neoleptoneta myopica), Kretschmarr Cave Mold Beetle (Texamaurops reddelli), and Tooth Cave Pseudoscorpion (Tartarocreagris texana) 5-year review: Summary and evaluation. Austin Ecological Services Field Office, Austin, TX. 13 pp.

Weygoldt, P. 1969. The biology of pseudoscorpions. Harvard University Press, Cambridge, MA. 160 pp.

CAVE HARVESTMEN

Goodnight, C. J., and M. L. Goodnight. 1967. Opilionids from Texas caves (Opiliones: Phalangodidae). American Museum Novitates 2301:1–8.

Reddell, J. R. The cave fauna of Texas with special reference to the western Edwards Plateau. Pp. 31–50 in The caves and karst of Texas (W. R. Elliott and G. Veni, eds.) National Speleological Society, Huntsville, AL. 252 pp.

Taylor, S. J., J. Krejca, J. E. Smith, et al. 2003. Investigations of the potential for Red Imported Fire Ant (Solenopsis invicta) impacts on rare karst invertebrates of Fort Hood, Texas. Illinois Natural History Survey, Center for Biodiversity Technical Report 2003 (9): 1–128.

Ubick, D., and T. S. Briggs. 1992. The harvestman family, Phalangodidae. 3. Revision of Texella Goodnight and Goodnight (Opiliones: Laniatores). Texas Memorial Museum, Speleological Monographs 3:155–240.

———. 2004. The harvestman family, Phalangodidae. 5. New records of Texella Goodnight and Goodnight (Opiliones: Laniatores). Texas Memorial Museum, Speleological Monographs 6:101–141.

CAVE MESHWEAVERS AND SPIDERS

Chamberlin, R. V., and W. J. Gertsch. 1958. The spider family Dictynidae in America north of Mexico. Bulletin of the American Museum of Natural History 116 (1): 1–152.

Cokendolpher, J. C. 2004a. A new Neoleptoneta spider from a cave at Camp Bullis, Bexar County, Texas (Araneae: Leptonetidae). Texas Memorial Museum, Speleological Monographs 6:63–69.

———. 2004b. Cicurina spiders from caves in Bexar County, Texas (Araneae: Dictynidae). Texas Memorial Museum, Speleological Monographs 6:13–58.

Gertsch, W. J. 1974. The spider family Leptonetidae in North America. Journal of Arachnology 1:145–203.

———. 1992. Distribution patterns and speciation

in North American cave spiders with a list of the troglobites and revision of the cicurinas of the subgenus *Cicurella*. Texas Memorial Museum, Speleological Monographs 3:75–122.

Ledford, J. M., D. Ubick, and J. C. Cokendolpher. 2005. Leptonetidae. Pp. 122–123 *in* Spiders of North America: An identification manual (D. Ubick, P. Paquin, P. E. Cushing, and V. Roth, eds.). American Arachnological Society. 377 pp.

Reddell, J. R. 1988. The subterranean fauna of Bexar County, Texas. Pp. 27–51 *in* The caves of Bexar County, 2nd ed. (G. Veni, ed.). Texas Memorial Museum, Speleological Monographs 2:1–300.

MOLD BEETLES

Chandler, D. S. 1992. The Pselaphidae (Coleoptera) of Texas caves. Texas Memorial Museum, Speleological Monographs 3:241–253.

Chandler, D. S., and J. R. Reddell. 2001. A review of the ant-like litter beetles found in Texas caves (Coleoptera: Staphylinidae: Pselaphinae). Texas Memorial Museum, Speleological Monographs 5:115–128.

Chandler, D. S., J. R. Reddell, and P. Paquin. 2009. New cave Pselaphinae and records from Texas, with a discussion of the relationships and distributions of the Texas troglobitic Pselaphinae (Coleoptera: Staphylinidae: Pselaphinae). Texas Memorial Museum, Speleological Monographs 7:125–140.

GROUND BEETLES

Barr, T. C., Jr. 1974. Revision of *Rhadine* LeConte (Coleoptera: Carabidae). 1. The *subterranea* group. American Museum Novitates 2539:1–30.

Mitchell, R. W. 1971a. Preference responses and tolerance of the troglobitic carabid beetle, *Rhadine subterranea*. International Journal of Speleology 3:261–267.

———. 1971b. Food and feeding habits of the troglobitic carabid beetle *Rhadine subterranea*. International Journal of Speleology 3:289–304.

Reddell, J. R. 1988. The subterranean fauna of Bexar County, Texas. Texas Memorial Museum, Speleological Monographs 2:27–51.

Reddell, J. R., and J. C. Cokendolpher. 2001. A new species of troglobitic *Rhadine* (Coleoptera: Carabidae) from Texas. Texas Memorial Museum, Speleological Monographs 5:109–114.

———. 2004. New species and records of cavernicole *Rhadine* (Coleoptera: Carabidae) from Camp Bullis, Texas. Texas Memorial Museum, Speleological Monographs 6:153–162.

Chapter Seven

TEXAS FRESHWATER AMPHIPODS

Boghici, R., and N. G. Van Broekhoven. 2001. Hydrogeology of the Rustler Aquifer, Trans-Pecos Texas. Pp. 207–225 *in* Aquifers of West Texas (R. E. Mace, W. F. Mullican III, and E. S. Angle, eds.). Texas Water Development Board Report 356:1–263.

Chowdhury, A. H., C. Ridgeway, and R. E. Mace. 2004. Origin of the waters in the San Solomon Spring system, Trans-Pecos Texas. Proceedings of the Aquifers of the Edwards Plateau Conference, San Angelo. Texas Water Development Board Report 260:315–344.

Cole, G. 1985. Analysis of the *Gammarus pecos* complex (Crustacea: Amphipoda) in Texas and New Mexico, USA. Journal of the Arizona-Nevada Academy of Science 20:93–104.

Cole, G. A. 1976. A new amphipod crustacean, *Gammarus hyalelloides* n. sp., from Texas. Transactions of the American Microscopy Society 95:80–85.

Cole, G. A., and E. L. Bousfield. 1970. A new freshwater *Gammarus* (Crustacea: Amphipoda) from western Texas. American Midland Naturalist 83:89–95.

Gervasio, V., D. Berg, B. Lang, et al. 2004. Genetic diversity in the *Gammarus pecos* species complex: Implications for conservation and regional biogeography in the Chihuahuan Desert. Limnology and Oceanography 49: 520–531.

Gibson, J. R., S. J. Harden, and J. N. Fries. 2008. Survey and distribution of invertebrates from selected springs of the Edwards Aquifer in Comal and Hays Counties, Texas. Southwestern Naturalist 53:74–84.

Guyton, W. F., and Associates. 1979. Geohydrology of Comal, San Marcos, and Hueco Springs. Texas Department of Water Resources Report 234:1–85.

Lang, B. K., V. Gervasio, D. J. Berg, et al. 2003. Gammarid amphipods of northern Chihuahuan Desert spring systems: An imperiled fauna. Pp. 47–57 *in* Aquatic fauna of the northern Chihuahuan Desert (G. P. Garrett and N. L. Allan, eds.). Museum of Texas Tech University Special Publications 46:1–160.

Mace, R. E., W. F. Mullican III, and E. S. Angle. 2001. Aquifers of Texas. Texas Water Development Board Report 356:1–263.

Seidel, R. A., B. K. Lang, and D. J. Berg. 2009. Phylogeographic analysis reveals multiple cryptic species of amphipods (Crustacea: Amphipoda) in Chihuahuan Desert springs. Biological Conservation 142:2303–2313.

———. 2010. Salinity tolerance as a potential driver of ecological speciation in amphipods (*Grammarus* spp.) from the northern Chihuahuan Desert. Journal of the North American Benthological Society 29:1161–1169.

COMAL SPRINGS AQUATIC BEETLES

Barr, C. B. 1993. Survey for two Edwards Aquifer invertebrates: Comal Springs Dryopid Beetle *Stygoparnus comalensis* Barr and Spangler (Coleoptera:Dryopidae) and Peck's Cave Amphipod *Stygobromus pecki* Holsinger (Amphipoda:Crangonyctidae). US Fish and Wildlife Service, Austin, TX.

Barr, C. B., and P. J. Spangler. 1992. A new genus and species of stygobiontic dryopid beetle, *Stygoparnus comalensis* (Coleoptera: Dryopidae), from Comal Springs, Texas. Proceedings of the Biological Society of Washington 105:40–54.

Bosse, L. S., D. W. Tuff, and H. P. Brown. 1988. A new species of *Heterelmis* from Texas (Coleoptera: Elmidae). Southwestern Naturalist 33:199–203.

Bowles, D. E., C. B. Barr, and R. Stanford. 2003. Habitat and phenology of the endangered riffle beetle *Heterelmis comalensis* and a coexisting species, *Microcylloepus pusillus*, (Coleoptera: Elmidae) at Comal Springs, Texas. Archiv für Hydrobiologie 156:361–383.

Cooke, M., G. Longley, and R. Gibson. 2015. Spring association and microhabitat preferences of the Comal Springs Riffle Beetle (*Heterelmis comalensis*). Southwestern Naturalist 60:110–121.

Diaz, P. H., and M. L. Alexander. 2012. Aquatic macroinvertebrates of a spring-fed ecosystem in Hays County, Texas, USA. Entomological News 121:478–486.

Fries, J. N. 2003. Possible reproduction of the Comal Springs Riffle Beetle, *Heterelmis comalensis* (Coleoptera: Elmidae), in captivity. Entomological News 114:7–9.

Gibson, J. R., S. J. Harden, and J. N. Fries. 2008. Survey and distribution of invertebrates from selected springs of the Edwards Aquifer in Comal and Hays Counties, Texas. Southwestern Naturalist 53:74–84.

Huston, D. C., and J. R. Gibson. 2015. Underwater pupation by the Comal Springs Riffle Beetle, *Heterelmis comalensis* Bosse, Tuff, and Brown, 1988 (Coleoptera: Elmidae), with an update on culture techniques. Coleopterists Bulletin 69:521–524.

Huston, D. C., J. R. Gibson, K. G. Ostrand, et al. 2015. Monitoring and marking techniques for the endangered Comal Springs Riffle Beetle, *Heterelmis comalensis* Bosse, Tuff, and Brown, 1988 (Coleoptera: Elmidae). Coleopterists Bulletin 69:793–798.

WEST TEXAS SPRINGSNAILS

Boghici, R. 1997. Hydrogeological investigations at Diamond Y Springs and surrounding area, Pecos County, Texas. Unpublished master's thesis, University of Texas at Austin. 120 pp.

Bradstreet, J. 2012. Habitat associations and abundance estimates of native and exotic freshwater snails in a West Texas spring. Unpublished master's thesis, Texas Tech University, Lubbock. 103 pp.

Brown, K. M., B. Lang, and K. E. Perez. 2008. The conservation ecology of North American pleurocerid and hydrobiid gastropods. Journal of the North American Benthological Society 27:484–495.

Brune, G. 2002. Springs of Texas. Vol. 1. 2nd ed. Texas A&M University Press, College Station. 566 pp.

Hershler, R., H.-P. Liu, and J. Howard. 2014. Springsnails: A new conservation focus in western North America. BioScience 64:693–700.

Hershler, R., H.-P. Liu, and M. Mulvey. 1999. Phylogenetic relationships within the aquatic snail genus *Tryonia*: Implications for biogeography of the North American Southwest. Molecular Phylogenetics and Evolution 13: 377–391.

Kabat, A. R., and R. Hershler. 1993. The prosobranch snail family Hydrobiidae (Gastropoda: Rissooidea): Review of classification and supraspecific taxa. Smithsonian Contributions to Zoology 547. 94 pp.

Lang, B. K. 2011. Population monitoring (2001–2009) of federal candidate hydrobiid snails and gammarid amphipods of West Texas: A report to the US Fish and Wildlife Service (Austin, TX). 13 pp.

Ridgeway, C. 2005. Diminished spring flows in the San Solomon Springs system, Trans-Pecos, Texas. Final report for grant # section 6 E-19. Submitted May 2005. Texas Parks and Wildlife Department, Austin. 121 pp.

Uliana, M. M., and J. M. Sharp Jr. 2001. Tracing regional flow paths to major springs in Trans-Pecos Texas using geochemical data and geochemical models. Chemical Geology 179:53–72.

Veni, G., and Associates. 1991. Delineation and preliminary hydrogeologic investigation of the Diamond Y Spring, Pecos County, Texas. Final report prepared for the Texas Nature Conservancy, San Antonio. 111 pp.

White, W. N., H. S. Gale, and S. S. Nye. 1977. Geology and ground-water resources of the Balmorhea area, western Texas. US Geological Survey Water-Supply Paper 849-C:83–146.

WEST TEXAS PUPFISHES

Black, A. N., J. L. Snekser, L. Al-Shaer, et al. 2016. A review of the Leon Springs Pupfish (*Cyprinodon bovinus*) long-term conservation strategy and response to habitat restoration. Aquatic Conservation: Marine and Freshwater Ecosystems 26:410–416.

Childs, M. R., A. A. Echelle, and T. E. Dowling. 1996.

Development of the hybrid swarm between Pecos Pupfish (Cyprinodontidae: *Cyprinodon pecosensis*) and Sheepshead Minnow (*Cyprinodon variegatus*): A perspective from allozymes and mtDNA. Evolution 50:2014–2022.

Echelle, A. A. and A. F. Echelle. 1997. Genetic introgression of endemic taxa by non-natives: A case study with Leon Springs Pupfish and Sheepshead Minnow. Conservation Biology 11:153–161.

Echelle, A. A., A. F. Echelle, S. Contreras-B., and M. L. Lozano-V. 2003. Pupfishes of the northern Chihuahuan Desert: Status and conservation. *In* Aquatic fauna of the northern Chihuahuan Desert (G. P. Garrett and N. L. Allan, eds.). Museum of Texas Tech University, Special Publications 46:111–126.

Garrett, G. P. 2003. Innovative approaches to recover endangered species. *In* Aquatic fauna of the northern Chihuahuan Desert (G. P. Garrett and N. L. Allan, eds.). Museum of Texas Tech University, Special Publications 46:151–160.

Karges, J. 2003. Aquatic conservation and The Nature Conservancy of West Texas. *In* Aquatic fauna of the northern Chihuahuan Desert (G. P. Garrett and N. L. Allan, eds.). Museum of Texas Tech University, Special Publications 46:141–150.

Kennedy, S. E. 1977. Life history of the Leon Springs Pupfish, *Cyprinodon bovinus*. Copeia 1977:93–103.

Leiser, J. K., and M. Itzkowitz. 2002. The relative costs and benefits of territorial defense and the two conditional male mating tactics in the Comanche Springs Pupfish (*Cyprinodon elegans*). Acta Ethologica 5:65–72.

US Fish and Wildlife Service. 2013a. Comanche Springs Pupfish (*Cyprinodon elegans*) 5-year review: Summary and evaluation. US Fish and Wildlife Service Field Office, Austin, TX. 39 pp.

———. 2013b. Leon Springs Pupfish (*Cyprinodon bovinus*) 5-year review: Summary and evaluation. US Fish and Wildlife Service Field Office, Austin, TX. 29 pp.

Winemiller, K. O., and A. A. Anderson. 1997. Response of endangered desert fish populations to a constructed refuge. Restoration Ecology 5:204–213.

CENTRAL TEXAS SALAMANDERS

Bendik, N. 2011. Jollyville Plateau Salamander status report. City of Austin Watershed Protection SR-11-10. 43 pp.

Bishop, S. C. 1941. Notes on salamanders with descriptions of several new forms. Occasional Papers of the Museum of Zoology, University of Michigan 451:6–9.

Bowles, B. D., M. S. Sanders, and R. S. Hansen. 2006. Ecology of the Jollyville Plateau Salamander (*Eurycea tonkawae*: Plethodontidae) with an assessment of the potential effects of urbanization. Hydrobiologia 553:111–120.

Chamberlain, D. A., and L. O'Donnell. 2003. City of Austin's captive breeding program for the Barton Springs and Austin Blind Salamanders (January 1–December 31, 2002). City of Austin Watershed Protection and Development Review Department annual permit report (PRT-839031).

Chippindale, P. T., A. H. Price, and D. M. Hillis. 1993. A new species of perennibranchiate salamander (*Eurycea*: Plethodontidae) from Austin, Texas. Herpetologica 49:248–259.

Chippindale, P. T., A. H. Price, J. J. Weins, and D. M. Hillis. 2000. Phylogenetic relationships and systematic revision of Central Texas hemidactyline plethodontid salamanders. Herpetological Monographs 14:1–80.

Cole, R. A. 1995. A review of status, research and management of taxon viability for three neotenic aquatic salamanders in Travis County, Texas. Pp. 15–57 *in* A review of the status of current critical biological and ecological information on the *Eurycea* salamanders located in Travis County, Texas (D. E. Bowles, ed). Resource Protection Division, Texas Parks and Wildlife Department, Austin. 94 pp.

Diaz, P. H., J. N. Fries, T. H. Bonner, M. L. Alexander, and W. H. Nowlin. 2015. Mesohabitat associations of the threatened San Marcos Salamander (*Eurycea nana*) across its geographic range. Aquatic Conservation: Marine and Freshwater Ecosystems 25:307–321.

Hillis, D. M., D. A. Chamberlain, T. P. Wilcox, and P. T. Chippindale. 2001. A new species of subterranean blind salamander (Plethodontidae: Hemidactyliini: *Eurycea*: *Typhlomolge*) from Austin, Texas, and nomenclature of the major clades of Central Texas paedomorphic salamanders. Herpetologica 53:266–280.

Najvar, P. A., J. N. Fries, and J. T. Baccus. 2007. Fecundity of San Marcos Salamanders in captivity. Southwestern Naturalist 52:145–147.

O'Donnell, L., M. Turner, E. Geismer, and M. Sanders. 2005. Summary of Jollyville Plateau Salamander data (1997–2005) and status. Prepared by City of Austin, Watershed Protection Department, Environmental Resources Management Division, Water Resource Evaluation Section. Austin, TX. 58 pp.

Pierce, B. A., J. L. Christiansen, A. L. Ritzer, and T. A. Jones. 2010. Distribution and ecology of the Georgetown Salamander, *Eurycea naufragia*. Southwestern Naturalist 55:296–301.

Pierce, B. A., K. D. McEntire, and A. E. Wall. 2014. Population size, movement, and reproduction of the Georgetown Salamander, *Eurycea naufragia*. Herpetological Conservation and Biology 9:137–145.

Pierce, B. A., and A. E. Wall. 2011. Review of research literature related to the biology, evolution, and conservation of Georgetown Salamander, *Eurycea naufragia*. Report to the Williamson County Conservation Foundation, Southwestern University, Georgetown, Texas. 40 pp.

Tupa, D. D., and W. K. Davis. 1976. Population dynamics of the San Marcos Salamander, *Eurycea nana* Bishop. Texas Journal of Science 27:179–195.

Chapter Eight

Brown, D. E., ed. 1983. The wolf in the southwest. University of Arizona Press, Tucson. 195 pp.

Wade, D. A., D. W. Hawthorne, G. L. Nunley, and M. Caroline. 1984. History and status of predator control in Texas. Proceedings of the Eleventh Vertebrate Pest Conference 42:122–131.

AMERICAN BURYING BEETLE

Anderson, R. S. 1982. On the decreasing abundance of *Nicrophorus americanus* Olivier (Coleoptera: Silphidae) in eastern North Americana. Coleopterists Bulletin 36:362–365.

Creighton, J. C., and G. D. Schnell. 1998. Short-term movement patterns of the endangered American Burying Beetle *Nicrophorus americanus*. Biological Conservation 86:281–287.

Creighton, J. C., C. C. Vaughn, and B. R. Chapman. 1993. Habitat preference of the American burying beetle (*Nicrophorus americanus*) in Oklahoma. Southwestern Naturalist 38:275–277.

Holloway, A. K., and G. D. Schnell. 1997. Relationship between numbers of the American burying beetle *Nicrophorus americanus* Olivier (Coleoptera: Silphidae) and available food resources. Biological Conservation 81:145–152.

Kozol, A. J., M. P. Scott, and J. F. A. Traniello. 1988. The American Burying Beetle, *Nicrophorus americanus*: Studies on the natural history of a declining species. Psyche 95:167–195.

Lomolino, M. V., and J. C. Creighton. 1996. Habitat selection, breeding success and conservation of the endangered American Burying Beetle, *Nicrophorus americanus*. Biological Conservation 77:235–241.

Lomolino, M. V., J. C. Creighton, G. D. Schnell, and D. L. Certain. 1995. Ecology and conservation of the endangered American Burying Beetle (*Nicrophorus americanus*). Conservation Biology 9:605–614.

Mullins, P. L., E. G. Riley, and J. D. Oswald. 2013. Identification, distribution, and adult phenology of the carrion beetles (Coleoptera: Silphidae) of Texas. Zootaxa 3666:221–251.

Scott, M. P. 1998. The ecology and behavior of burying beetles. Annual Review of Entomology 43:595–618.

Sikes, D. S., and R. J. Raithel. 2002. A review of hypotheses of decline of the endangered American Burying Beetle (Silphidae: *Nicrophorus americanus* Olivier). Journal of Insect Conservation 6:103–113.

Trumbo, S. T., and P. L. Bloch. 2000. Habitat fragmentation and burying beetle abundance and success. Journal of Insect Conservation 4:245–252.

US Fish and Wildlife Service. 2008. American Burying Beetle (*Nicrophorus americanus*) 5-year review: Summary and evaluation. New England Field Office, Concord, NH. 53 pp.

HOUSTON TOAD

Forstner, M. R. J., and J. Dixon. 2011. Houston Toad (*Bufo houstonensis*) 5-year review: Summary and evaluation. Final report for section 6 project E-101. Submitted to Texas Parks and Wildlife Department and US Fish and Wildlife Service. 64 pp.

Greuter, K. 2004. Early juvenile ecology of the endangered Houston Toad, *Bufo houstonensis* (Anura: Bufonidae). Unpublished master's thesis, Texas State University, San Marcos. 102 pp.

Hatfield, J. S., A. H. Price, D. D. Diamond, and C. D. True. 2004. Houston Toad (*Bufo houstonensis*) in Bastrop County, Texas: Need for protecting multiple populations. Pp. 292–298 *in* Species conservation and management: Case studies (H. R. Akçakaya, M. A. Burgman, O. Kindvall, et al., eds). Oxford University Press, New York. 552 pp.

Hillis, D. M., A. M. Hillis, and R. F. Martin. 1984. Reproductive ecology and hybridization of the endangered Houston Toad (*Bufo houstonensis*). Journal of Herpetology 18:56–71.

Kennedy, J. P. 1962. Spawning season and experimental hybridization of the Houston Toad, *Bufo houstonensis*. Herpetologica 17:239–245.

McHenry, D. J., and M. R. J. Forstner. 2009. Houston Toad metapopulation assessment and genetics: Data necessary for effective recovery strategies in a significantly fragmented landscape. Final report for section 6 project E-76. Submitted to Texas Parks and Wildlife Department and US Fish and Wildlife Service. 110 pp.

Sanders, O. 1953. A new species of toad, with a discussion of morphology of the bufonid skull. Herpetologica 9:25–47.

Swannack, T. M. 2007. Ecology of the Houston Toad (*Bufo houstonensis*). PhD diss., Texas A&M University, College Station. 154 pp.

LOUISIANA PINESNAKE

Conant, R. 1956. A review of two rare pine snakes from the Gulf Coastal Plain. American Museum Novitates 1781: 1–31.

Ealy, M. J., R. R. Fleet, and D. C. Rudolph. 2004. Diel activity patterns of the Louisiana Pine Snake (*Pituophis ruthveni*) in eastern Texas. Texas Journal of Science 56:383–394.

Himes, J. G., L. M. Hardy, D. C. Rudolph, and S. J. Burgdorf. 2006. Movement patterns and habitat selection by native and repatriated Louisiana Pine Snakes (*Pituophis ruthveni*): Implications for conservation. Herpetological Natural History 9:103–116.

Rudolph, D. C., and S. Burgdorf. 1997. Timber Rattlesnakes and Louisiana Pine Snakes of the west Gulf Coastal Plain: Hypotheses of decline. Texas Journal of Science 49:111–122.

Rudolph, D. C., S. J. Burgdorf, R. R. Schaefer, et al. 2006. Status of *Pituophis ruthveni* (Louisiana Pine Snake). Southeastern Naturalist 5:463–472.

Rudolph, D. C., R. N. Conner, R. R. Schaefer, and R. W. Maxey. 2005. Determination of the status of the Louisiana Pine Snake. Wildlife Research Highlights, Texas Parks and Wildlife Department 7:71–72.

Rudolph, D. C., C. A. Melder, J. Pierce, et al. 2012. Diet of the Louisiana Pine Snake (*Pituophis ruthveni*). Herpetological Review 43:243–245.

Rudolph, D. C., R. R. Schaefer, S. J. Burgdorf, et al. 2007. Pine snake (*Pituophis ruthveni* and *Pituophis melanoleucus lodingi*) hibernacula. Journal of Herpetology 41:560–565.

Thomas, R. A., B. J. Davis, and M. R. Culbertson. 1976. Notes on variation and home-range of the Louisiana Pine Snake, *Pituophis melanoleucus ruthveni* Stull (Reptilia, Serpentes, Colubridae). Journal of Herpetology 10:252–254.

Werler, J. E., and J. R. Dixon. 2000. Texas snakes: Identification, distribution and natural history. University of Texas Press, Austin. 437 pp.

MEXICAN LONG-NOSED BAT

Adams, E. R. 2015. Seasonal and nightly activity of Mexican Long-nosed Bats (*Leptonycteris nivalis*) in Big Bend National Park, Texas. Unpublished master's thesis, Angelo State University, San Angelo, TX. 64 pp.

Adams, E. R., and L. K. Ammerman. 2015. Serpentine antenna configuration for passive integrated transponder tag readers used at bat roosts. Southwestern Naturalist 60:393–397.

Ammerman, L. K., M. McDonough, N. I. Hristov, and T. H. Kunz. 2009. Census of the endangered Mexican Long-nosed Bat *Leptonycteris nivalis* in Texas, USA, using thermal imaging. Endangered Species Research 8:87–92.

Arita, H. T., and D. E. Wilson. 1987. Long-nosed bats and agaves: The tequila connection. Bats 5:3–5.

Brown, C. M. 2008. Natural history and population genetics of the endangered Mexican Long-nosed Bat, *Leptonycteris nivalis* (Chiroptera: Phyllostomidae). Unpublished master's thesis, Angelo State University, San Angelo, TX. 128 pp.

England, A. E. 2004. Radio-tracking of the Greater Long-nosed Bat, *Leptonycteris nivalis*, in Big Bend National Park, Texas. Bat Research News 40:95–96.

Hensley, A. P., and K. T. Wilkins. 1988. *Leptonycteris nivalis*. Mammalian Species 307:1–4.

Howell, D. J. 1979. Flock foraging in nectar-feeding bats: Advantages to the bats and to the host plants. American Naturalist 114:23–49.

Hoyt, R. A., J. S. Altenbach, and D. J. Hafner. 1994. Observations on the long-nosed bats (*Leptonycteris*) in New Mexico. Southwestern Naturalist 39:175–179.

Mollhagen, T. 1973. Distributional and taxonomic notes on some West Texas bats. Southwestern Naturalist 17: 427–430.

Moreno-Valdez, A., R. L. Honeycutt, and W. E. Grant. 2004. Colony dynamics of *Leptonycteris nivalis* (Mexican Long-nosed Bat) related to flowering *Agave* in northern Mexico. Journal of Mammalogy 85:453–459.

Sanchez, R., and R. A. Medellín. 2007. Food habits of the threatened bat *Leptonycteris nivalis* (Chiroptera: Phyllostomatidae) in a mating roost in Mexico. Journal of Natural History 41:1753–1764.

LOUISIANA BLACK BEAR

Kaminski, D. J., C. E. Comer, N. P. Garner, et al. 2013. Using GIS-based, landscape-scale suitability modeling to identify conservation priority areas: A case study of the Louisiana Black Bear (*Ursus americanus luteolus*) in East Texas. Journal of Wildlife Management 77:1639–1649.

Pelton, M. R. 2003. Black Bear. Pp. 547–555 *in* Wild mammals of North America: Biology, management, and conservation, 2nd ed. (G. A. Feldhamer, B. C. Thompson, and J. A. Chapman, eds.). Johns Hopkins University Press, Baltimore, MD. 1232 pp.

Texas Parks and Wildlife Department. 2005. East Texas

Black Bear conservation and management plan 2005–
2015. Austin. 56 pp.

Triant, D. A., R. M. Pace, and M. Stine. 2004. Abundance,
genetic diversity and conservation of Louisiana Black
Bears as detected through noninvasive sampling.
Conservation Genetics 5:647–659.

US Fish and Wildlife Service. 2016. Post-delisting
monitoring plan for the Louisiana Black Bear (*Ursus
americanus luteolus*). US Fish and Wildlife Service,
Lafayette, LA. 52 pp.

MEXICAN GRAY WOLF

Bogan, M. A., and P. Melhop. 1983. Systematic relation-
ships of Gray Wolves (*Canis lupus*) in southwestern
North America. Occasional Papers, Museum of
Southwestern Biology 1:1–21.

Holaday, B. 2003. Return of the Mexican Gray Wolf: Back
to the blue. University of Arizona Press, Tempe. 270 pp.

Mech, L. D. 1974. *Canis lupus*. Mammalian Species 37:1–6.

Reed, J. E., W. B. Ballard, P. S. Gipson, et al. 2009. Diets of
free-ranging Mexican Gray Wolves in Arizona and New
Mexico. Wildlife Society Bulletin 34:1127–1133.

RED WOLF

Beeland, T. D. 2013. The secret world of Red Wolves: The
fight to save North America's other wolf. University of
North Carolina Press, Chapel Hill. 272 pp.

Hinton, J. W., M. J. Chamberlain, and D. R. Rabon Jr.
2013. Red Wolf (*Canis rufus*) recovery: A review with
suggestions for future research. Animals 3:722–724.

Nowak, R. M. 1992. The Red Wolf is not a hybrid.
Conservation Biology 6:593–595.

Paradiso, J. L. 1968. Canids recently collected in East
Texas, with comments on the taxonomy of the Red
Wolf. American Midland Naturalist 80:529–534.

Paradiso, J. L., and R. M. Nowak. 1972. *Canis rufus*.
Mammalian Species 22:1–4.

Rutledge, L. Y., S. Devillard, J. Q. Boone, et al. 2015. RAD
sequencing and genomic simulations resolve hybrid
origins within North American *Canis*. Biology Letters
11:1–4.

JAGUAR

Bailey, V. 1905. Biological survey of Texas. North American
Fauna 25. Government Printing Office, Washington,
DC. 222 pp.

Brown, D. E. 1983. On the status of the Jaguar in the
Southwest. Southwestern Naturalist 28:459–460.

Brown, D. E., and C. A. Lopez-Gonzalez. 2001. Borderland
Jaguars: *Tigres de la frontera*. University of Utah Press,
Salt Lake City. 170 pp.

McCain, E. B., and J. L. Childs. 2008. Evidence of resident
Jaguars (*Panthera onca*) in the southwestern United
States and the implications for conservation. Journal of
Mammalogy 89:1–10.

Sanderson, E. W., K. H. Redford, C. B. Chetkiewicz,
et al. 2002. Planning to save a species: The Jaguar as a
model. Conservation Biology 16:58–72.

Seymour, K. L. 1989. *Panthera onca*. Mammalian Species
340:1–9.

GULF COAST JAGUARUNDI

Cain, A. T., V. R. Tuovila, D. G. Hewitt, and M. E. Tewes.
2003. Effects of a highway and mitigation projects on
Bobcats in southern Texas. Biological Conservation 114:
189–197.

de Oliveira, T. G. 1998. *Herpailurus yagouaroundi*.
Mammalian Species 578:1–6.

Manzani, P. R., and F. Monteiro. 1989. Notes on the
food habits of the jaguarundi, *Felis yagouaroundi*
(Mammalia: Carnivora). Mammalia 53:659–660.

McCarthy, T. J. 1992. Notes concerning the jaguarundi cat
(*Herpailurus yagouaroundi*) in the Caribbean lowlands
of Belize and Guatemala. Mammalia 56:302–306.

Tewes, M. E., and D. D. Everett. 1986. Status and
distribution of the endangered Ocelot and Jaguarundi
in Texas. Pp. 147–158 *in* Cats of the world: Biology,
conservation and management (S. D. Miller and
D. D. Everett, eds.). National Wildlife Federation,
Washington, DC. 501 pp.

Tewes, M. E., and D. J. Schmidly. 1987. The Neotropical
felids: Jaguar, Ocelot, Margay, and Jaguarundi. Pp. 697–
711 *in* Wild furbearer management and conservation in
North America (M. Nowak, J. A. Baker, M. E. Obbard,
and B. Malloch, eds.). Ontario Ministry of Natural
Resources, Concord, Ontario, Canada. 1150 pp.

OCELOT

Haines, A. M., J. E. Janecka, M. E. Tewes, et al. 2006. The
importance of private lands for Ocelot *Leopardus
pardalis* conservation in the United States. Oryx 40:
90–94.

Haines, A. M., M. E. Tewes, and L. L. Laack. 2005. Survival
and sources of mortality in Ocelots. Journal of Wildlife
Management 69:255–263.

Haines, A. M., M. E. Tewes, L. L. Laack, et al. 2006. Habitat
based population viability analysis of Ocelots in
southern Texas. Biological Conservation 132:424–436.

Harveson, P. M., M. E. Tewes, G. L. Anderson, and L. L.
Laack. 2004. Habitat use by Ocelots in South Texas:
Implications for restoration. Wildlife Society Bulletin
32:948–954.

Janecka, J. E., M. E. Tewes, L. L. Laack, et al. 2011. Reduced genetic diversity and isolation of remnant Ocelot populations occupying a severely fragmented landscape in southern Texas. Animal Conservation 14: 608–619.

Laack, L. L., M. E. Tewes, A. M. Haines, and J. Rappole. 2005. Reproductive life history of Ocelots *Leopardus pardalis* in southern Texas. Acta Theriologica 50: 505–514.

Shindle, D. B., and M. E. Tewes. 1998. Woody species composition of habitats used by Ocelots (*Leopardus pardalis*) in the Tamaulipan Biotic Province. Southwestern Naturalist 43:273–279.

Tewes, M. E., and M. G. Hornocker. 2008. Effects of drought on Bobcats and Ocelots. Pp. 123–138 *in* Wildlife science: Linking ecological theory to management applications (T. Fulbright and D. Hewitt, eds.). CRC Press, Boca Raton, FL. 384 pp.

Tewes, M. E., L. L. Laack, and A. Caso. 1995. Corridor management for Ocelots in the United States and northern Mexico. Pp. 444–446 *in* Integrating people and wildlife for a sustainable future (J. A. Bissonette and P. R. Krausman, eds.). Proceedings of the First International Wildlife Management Congress. The Wildlife Society, Bethesda, MD. 715 pp.

Chapter Nine

King, K. A., and E. L. Flickinger. 1977. The decline of the Brown Pelican on the Louisiana and Texas Gulf Coast. Southwestern Naturalist 21:417–431.

Robbins, C. S., J. R. Sauer, R. S. Greenberg, and S. Droeg. 1989. Population declines in North American birds that migrate to the Neotropics. Proceedings of the National Academy of Science 86:7658–7662.

Schreiber, R. W. 1980. The Brown Pelican: An endangered species? Bioscience 30:742–747.

Terborgh, J. 1989. Where have all the birds gone? Princeton University Press, Princeton, NJ. 224 pp.

WHOOPING CRANE

Allen, R. P. 1952. The Whooping Crane. National Audubon Society Resource Report 3. 246 pp.

Blankinship, D. R. 1976. Studies of Whooping Cranes on the wintering grounds. Pp. 197–206 *in* Proceedings of the International Crane Workshop (J. C. Lewis, ed.). Oklahoma State University, Stillwater. 355 pp.

Gil-Weir, K. C., W. E. Grant, R. D. Slack, et al. 2012. Demography and population trends of Whooping Cranes. Journal of Field Ornithology 83:1–10.

Hunt, H. E., and R. D. Slack. 1989. Winter diets of Whooping and Sandhill Cranes in South Texas. Journal of Wildlife Management 53:1150–1154.

Kuyt, E., and P. J. Goossen. 1987. Survival, age composition, sex ratio, and age at first breeding of Whooping Cranes in Wood Buffalo National Park, Canada. Pp. 230–244 *in* Proceedings of the 1985 Crane Workshop (J. C. Lewis, ed.). Platte River Whooping Crane Habitat Maintenance Trust and US Fish and Wildlife Service, Grand Island, NE. 415 pp.

Lewis, J. C., E. Kuyt, K. E. Schwindt, and T. V. Stehn. 1992. Mortality in fledged cranes of the Aransas-Wood Buffalo population. Pp. 145–148 *in* Proceedings of the 1988 North American Crane Workshop (D. A. Wood, ed.). State of Florida Game and Fresh Water Fish Commission, Lake Wales. 305 pp.

Pugesek, B. H., M. J. Baldwin, and T. Stehn. 2013. The relationship of Blue Crab abundance to winter mortality of Whooping Cranes. Wilson Journal of Ornithology 125:658–661.

Stehn, T. V., and E. F. Johnson. 1987. The distribution of winter territories of the Whooping Crane on the Texas coast. Pp. 180–195 *in* Proceedings of the 1985 Crane Workshop (J. C. Lewis, ed.). Platte River Whooping Crane Habitat Maintenance Trust and US Fish and Wildlife Service, Grand Island, NE. 415 pp.

Timoney, K. P. 1999. The habitat of nesting Whooping Cranes. Biological Conservation 89:189–197.

Wozniak, J. R., T. M. Swannack, R. Butzler, et al. 2012. River inflow, estuarine salinity, and Carolina Wolfberry fruit abundance: Linking abiotic drivers to Whooping Crane food. Journal of Coastal Conservation 16:345–354.

Wright, G. D., M. J. Harner, and J. D. Chambers. 2014. Unusual wintering distribution and migratory behavior of the Whooping Crane (*Grus americana*) in 2011–2012. Wilson Journal of Ornithology 126:115–120.

ESKIMO CURLEW

Baird, S. F., T. M. Brewer, and R. Ridgway. 1884. The water birds of North America. Vol. 1. Little, Brown, Boston, MA.

Bent, A. C. 1928. Life histories of North American shore birds, part 1. Smithsonian Institution National Museum Bulletin 142. Washington, DC. 412 pp.

Blankinship, D. R., and K. A. King. 1984. A probable sighting of 23 Eskimo Curlews in Texas. American Birds 38 (6): 1066–1067.

Gollop, J. B., ed. 1986. Eskimo Curlew: A vanishing species? Saskatchewan Natural History Society, Regina. 159 pp.

Swenk, M. H. 1916. The Eskimo Curlew and its disappearance. Pp. 325–340 *in* Annual Report of the Smithsonian Institution for 1915. Smithsonian Institution, Washington, DC. 705 pp.

———. 1926. The Eskimo Curlew in Nebraska. Wilson Bulletin 38:117–118.

Williams, G. G. 1959. Probable Eskimo Curlew on Galveston Island, Texas. Auk 76:539–541.

WESTERN ATLANTIC RED KNOT

Baker, A., P. Gonzalez, R. I. G. Morrison, and Brian A. Harrington. 2013. Red Knot (*Calidris canutus*). *In* The birds of North America online (A. Poole, ed.). Cornell Laboratory of Ornithology, Ithaca, NY. http://bna.birds .cornell.edu/bna/species/563; doi:10.2173/bna.563.

Baker, A. J., P. M. González, T. Piersma, et al. 2004. Rapid population decline in Red Knots: Fitness consequences of decreased refueling rates and late arrival in Delaware Bay. Proceedings of the Royal Society B–Biological Sciences 271 (1541): 875–882.

Burger, J., L. J. Niles, R. R. Porter, et al. 2012. Migration and over-wintering of Red Knots (*Calidris canutus rufa*) along the Atlantic coast of the United States. Condor 114:302–313.

Harrington, B. 1996. The flight of the Red Knot. W. A. Norton, New York. 192 pp.

Newstead, D. J., L. J. Niles, R. R. Porter, et al. 2013. Geolocation reveals mid-continent migratory routes and Texas wintering areas of Red Knots *Calidris canutus rufa*. Wader Study Group Bulletin 120:53–59.

Niles, L. J., H. P. Sitters, A. D. Dey, et al. 2008. Status of the Red Knot (*Calidris canutus rufa*) in the Western Hemisphere. Studies in Avian Biology 36:1–204.

Niles, L., H. Sitters, A. Dey, and the Red Knot Status Assessment Group. 2010. Red Knot conservation plan for the Western Hemisphere (*Calidris canutus*). Western Hemisphere Shorebird Reserve Network, Manomet, MA. 173 pp.

PIPING PLOVER

Elliott-Smith, E., M. Bidwell, A. E. Holland, and S. M. Haig. 2015. Data from the 2011 International Piping Plover Census. US Geological Survey Data Series 922. Reston, VA. http://dx.doi.org/10.3133/ds922. 296 pp.

Elliott-Smith, E., and S. M. Haig. 2004. Piping Plover (*Charadrius melodus*). *In* The birds of North America online (A. Poole, ed.). Cornell Laboratory of Ornithology, Ithaca, NY. http://bna.birds.cornell.edu /bna/species/002; doi:10.2173/bna.2.

Haig, S. M. 1987. Winter distribution and status of Piping Plovers in the Gulf of Mexico. Pp. 251–258 *in* Proceedings of the Workshop on Endangered Species in the Prairie Provinces (G. L. Holroyd., W. B.

McGillivray, P. H. R. Stepney, et al., eds.). Provincial Museum of Alberta, Natural History Occasional Paper No. 9. 367 pp.

INTERIOR LEAST TERN

Boyd, R. L., and B. C. Thompson. 1985. Evidence for reproductive mixing of Least Tern populations. Journal of Field Ornithology 56:405–406.

Boylen, J., O. Bocanegra, A. C. Kasner, and J. Beall. 2004. Use of non-traditional habitats by Interior Least Terns in north-central Texas. Pp. 105–117 *in* Third Missouri River and North American Piping Plover and Least Tern Habitat Workshop/Symposium (K. F. Higgins, M. R. Brashier, and C. Fleming, eds.). South Dakota State University, Brookings. 129 pp.

Burger, J. 1884. Colony stability in Least Terns. Condor 86: 61–67.

Butcher, J. A., R. L. Neill, and J. T. Boylan. 207. Survival of Interior Least Tern chicks hatched on gravel-covered roofs in North Texas. Waterbirds 30:595–601.

Kasner, A. C. 2004. Nesting and foraging ecology of Interior Least Terns nesting on reclaimed mine spoil. PhD diss., Texas A&M University, College Station.

Kasner, A. C., T. C. Maxwell, and R. D. Slack. 2005. Breeding distributions of selected Charadriiforms (Charadriiformes: Charadriidae, Scolopacidae, Laridae) in Texas. Texas Journal of Science 57:273–288.

Renken, R. B., and J. W. Smith. 1995. Interior Least Tern site fidelity and dispersal. Colonial Waterbirds 18:193–198.

Sidle, J. G., D. E. Carlson, E. M. Kirsch, and J. J. Dinan. 1992. Flooding: Mortality and habitat renewal for Least Terns and Piping Plovers. Colonial Waterbirds 15: 132–136.

Thompson, B. C., J. A. Jackson, J. Burger, et al. 1997. Least Tern (*Sterna antillarum*). *In* The birds of North America, No. 290 (A. Poole and F. Gill, eds.). Academy of Natural Sciences, Philadelphia, PA, and American Ornithologists' Union, Washington, DC.

US Fish and Wildlife Service. 2013. Interior Least Tern (*Sternula antillarum*) 5-year review: Summary and evaluation. US Fish and Wildlife Service, Jackson, MS. 71 pp.

PRAIRIE-CHICKENS

Bent, A. C. 1932. Life histories of North American gallinaceous birds. US National Museum Bulletin 162. Smithsonian Institution, Washington, DC. 490 pp.

Hagen, C. A., and K. M. Giesen. 2005. Lesser Prairie-Chicken (*Tympanuchus pallidicinctus*). *In* The birds of North America online (P. G. Rodewald, ed.). Cornell Laboratory of Ornithology, Ithaca, NY. https://birdsna

.org/Species-Account/bna/species/lepchi; doi:10.2173/bna.364.

Hagen, C. A., B. E. Jamison, K. M. Giesen, and T. Z. Riley. 2004. Guidelines for managing Lesser Prairie-Chicken populations and their habitats. Wildlife Society Bulletin 32:69–82.

Hammerstrom, F., and F. Hammerstrom. 1961. Status and problems of North American grouse. Wilson Bulletin 73:284–294.

Haukos, D. A., and C. W. Boal, eds. 2016. Ecology and conservation of Lesser Prairie Chickens. Studies in Avian Biology No. 48. CRC Press, Boca Raton, FL. 376 pp.

Horton, R., L. Bell, C. M. O'Meilia, et al. 2010. A spatially-based planning tool designed to reduce negative effects of development on the Lesser Prairie-Chicken (*Tympanuchus pallidicinctus*) in Oklahoma: A multi-entity collaboration to promote Lesser Prairie-Chicken voluntary habitat conservation and prioritized management actions. Oklahoma Department of Wildlife Conservation, Oklahoma City. 79 pp.

Jones, R. E. 1964. The specific distinctness of the Greater and Lesser Prairie-Chickens. Auk 81:65–73.

Lehmann, V. W. 1939. Heath Hen of the south. American Wildlife 28 (5): 221–227.

Sandercock, B. K., K. Martin, and G. Segelbacher, eds. 2011. Ecology, conservation, and management of grouse. Studies in Avian Biology No. 39. University of California Press, Berkeley. 376 pp.

MEXICAN SPOTTED OWL

Gutiérrez, R. J., A. B. Franklin, and W. S. Lahaye. 1995. Spotted Owl (*Strix occidentalis*). *In* The birds of North America online, No. 179 (A. Poole, ed.). Cornell Laboratory of Ornithology, Ithaca, NY. http://bna.birds.cornell.edu/bna/species/179; doi:10.2173/bna.179.

Mullet, T. C., and J. P. Ward Jr. 2010. Microhabitat features at Mexican Spotted Owl nest and roost sites in the Guadalupe Mountains. Journal of Raptor Research 44:277–285.

NORTHERN APLOMADO FALCON

Brown, J. L., and M. W. Collopy. 2008. Nest-site characteristics affect daily nest-survival rates of Northern Aplomado Falcons (*Falco femoralis septentrionalis*). Auk 125:105–112.

Brown, J. L., A. B. Montoya, E. J. Gott, and M. Curti. 2003. Piracy as an important foraging method of Aplomado Falcons in southern Texas and northern Mexico. Wilson Bulletin 115:357–359.

Cade, T. J. 1982. The falcons of the world. Cornell University Press, Ithaca, NY. 192 pp.

Cade, T. J., J. P. Penny, and B. J. Walton. 1991. Efforts to restore the Northern Aplomado Falcon *Falco femoralis septentrionalis* by captive breeding and reintroduction. Dodo 27:71–81.

Hector, D. P. 1986. Cooperative hunting and its relationship to foraging success and prey size in an avian predator. Ethology 73:247–257.

———. 1987. The decline of the Aplomado Falcon in the United States. American Birds 41:381–389.

Hunt, W. G., J. L. Brown, T. J. Cade, et al. 2013. Restoring Aplomado Falcons to the United States. Journal of Raptor Research 47:335–351.

Jenny, J. P., W. Heinrich, A. B. Montoya, et al. 2004. Progress in restoring the Aplomado Falcon to southern Texas. Wildlife Society Bulletin 32:276–285.

Mora, M. A., M. C. Lee, J. P. Penny, et al. 1997. Potential effects of environmental contaminants on recovery of the Aplomado Falcon in South Texas. Journal of Wildlife Management 61:1288–1296.

Perez, C. J., P. J. Zwank, and D. W. Smith. 1996. Survival, movements, and habitat use of Aplomado Falcons released in southern Texas. Journal of Raptor Research 30:175–182.

Truett, J. C. 2002. Aplomado Falcons and grazing: Invoking history to plan restoration. Southwestern Naturalist 47:379–400.

US Fish and Wildlife Service. 2014. Northern Aplomado Falcon (*Falco femoralis septentrionalis*) 5-year review: Summary and evaluation. US Fish and Wildlife Service, New Mexico Ecological Services Field Office, Albuquerque.

SPRAGUE'S PIPIT

Davis, S. K., D. C. Duncan, and M. Skeel. Distribution and habitat associations of three endemic grassland songbirds in southern Saskatchewan. Wilson Bulletin 111:389–396.

Emlen, J. T. 1972. Size and structure of a wintering avian community in southern Texas. Ecology 53:317–329.

Grzybowski, J. A. 1982. Population structure in grassland bird communities during winter. Condor 84:137–152.

Igl, L. D., and B. M. Ballard. 1999. Habitat associations of migrating and overwintering grassland birds in southern Texas. Condor 101:771–782.

Robbins, M. B., and B. C. Dale. 1999. Sprague's Pipit (*Anthus spragueii*). *In* The birds of North America, No. 489 (A. Poole and F. Gill, eds.). Academy of Natural Sciences, Philadelphia, PA, and American Ornithologists' Union, Washington, DC.

Winter, M., D. H. Johnson, J. A. Shaffer, et al. 2006. Patch size and landscape effects on density and nesting success of grassland birds. Journal of Wildlife Management 70:158–172.

SOUTHWESTERN WILLOW FLYCATCHER

Finch, D. M., and S. H. Stoleson, eds. 2000. Status, ecology, and conservation of the Southwestern Willow Flycatcher. USDA Forest Service General Technical Report RMRS-GTR-60. Rocky Mountain Research Station, Ogden, UT. 132 pp.

McCabe, R. A. 1991. The little green bird: Ecology of the Willow Flycatcher. Palmer Publications, Amherst, WI.

Oberholser, H. C. 1974. The bird life of Texas. University of Texas Press, Austin. 1069 pp.

Sedgwick, J. A. 2000. Willow Flycatcher (*Empidonax traillii*). *In* The birds of North America, No. 533 (A. Poole and F. Gill, eds.). Academy of Natural Sciences, Philadelphia, PA, and American Ornithologists' Union, Washington, DC.

Sogge, M. K., D. Ahlers, and S. J. Sferra. 2010. A natural history summary and survey protocol for the Southwestern Willow Flycatcher. US Geological Survey Techniques and Methods 2A-10. Reston, VA. 210 pp.

Sogge, M. K., B. E. Kus, S. J. Sferra, and M. J. Whitfield, eds. 2000. Ecology and conservation of the Willow Flycatcher. Studies in Avian Biology 26:1–210.

Zink, R. M. 2015. Genetics, morphology, and ecological niche modeling do not support the subspecies status of the endangered Southwestern Willow Flycatcher (*Empidonax traillii extimu*s). Condor 117:76–86.

WESTERN YELLOW-BILLED CUCKOO

Franzreb, K. E., and S. A. Laymon. 1993. A reassessment of the taxonomic status of the Yellow-billed Cuckoo. Western Birds 24:17–28.

Gaines, D. A., and S. A. Laymon. 1984. Decline, status and preservation of the Yellow-billed Cuckoo in California. Western Birds 15:49–80.

Halterman, M. D. 2009. Sexual dimorphism, home range, detection probability, and parental care in the Yellow-billed Cuckoo. PhD diss., University of Nevada, Reno.

Hamilton, W. J., III, and M. E. Hamilton. 1965. Breeding characteristics of the Yellow-billed Cuckoo in Arizona. Proceedings of the California Academy of Sciences, 4th Series 32:405–432.

Johnson, M. J., J. R. Hatten, J. A. Holmes, and P. B. Shafroth. 2017. Identifying Western Yellow-billed Cuckoo breeding habitat with a dual modelling approach. Ecological Modelling 347:50–62.

Johnson, M. J., J. A. Holmes, C. Calvo, et al. 2007. Yellow-billed Cuckoo distribution, abundance, and habitat use along the lower Colorado and tributaries, 2006 annual report. US Geological Survey Open File Report 2007-1097. 219 pp.

Laymon, S. A., and M. D. Halterman. 1987. Can the western subspecies of Yellow-billed Cuckoo be saved from extinction? Western Birds 18:19–25.

Sechrist, J., D. D. Ahlers, K. P. Zehfuss, et al. 2013. Home range and use of habitat of Western Yellow-billed Cuckoos on the Middle Rio Grande, New Mexico. Southwestern Naturalist 58:411–419.

Sechrist, J. D., E. H. Paxton, D. D. Ahlers, et al. 2012. One year of migration data for a Western Yellow-billed Cuckoo. Western Birds 43:2–11.

BLACK-CAPPED VIREO

Grzybowski, J. A. 1995. Black-capped vireo (*Vireo atricapillus*). *In* The birds of North America, No. 181 (A. Poole and F. Gill, eds.). Academy of Natural Sciences, Philadelphia, PA, and American Ornithologists' Union, Washington, DC.

Kostecke, R. M., S. G. Summers, G. H. Eckrich, and D. A. Cimprich. 2005. Effects of Brown-headed Cowbird (*Molothrus ater*) removal on Black-Capped Vireo (*Vireo atricapilla*) nest success and population growth at Fort Hood, Texas. Ornithological Monographs 57:28–37.

US Fish and Wildlife Service. 2016. Species status assessment report for the Black-capped Vireo (*Vireo atricapilla*). US Fish and Wildlife Service, Arlington, TX. 138 pp.

GOLDEN-CHEEKED WARBLER

Bolsinger, J. S. 2000. Use of two song categories by Golden-cheeked Warblers. Condor 102:539–552.

Eisermann, K., and U. Schulz. 2005. Birds of a high-altitude cloud forest in Alta Verapaz, Guatemala. Revista de Biología Tropical 53:577–594.

Groce, J. E., K. N. Smith, R. N. Wilkins, and D. Wolfe. 2012. The Golden-cheeked Warbler: History of a conflict. Wildlife Society Bulletin 36:401–407.

Ladd, C., and L. Gass. 1999. Golden-cheeked Warbler (*Setophaga chrysoparia*). *In* The birds of North America online (P. G. Rodewald, ed.). Cornell Laboratory of Ornithology, Ithaca, NY. https://birdsna.org/Species -Account/bna/species/gchwar; doi:10.2173/bna.420.

Lockwood, M. W. 1996. Courtship behavior of Golden-cheeked Warblers. Wilson Bulletin 108:591–592.

Pulich, W. M. 1976. The Golden-cheeked Warbler. Texas Parks and Wildlife Department, Austin. 172 pp.

Rappole, J. H., D. I. King, and W. C. Barrow Jr. 1999. Winter ecology of the endangered Golden-cheeked Warbler. Condor 101:762–770.

Rappole, J. H., D. I. King, and J. Dietz. 2003. Winter-vs. breeding habitat limitation for an endangered avian migrant. Ecological Applications 13:735–742.

RED-COCKADED WOODPECKER

Conner, R. N., and D. C. Rudolph. 1991. Forest habitat loss, fragmentation, and Red-cockaded Woodpeckers. Wilson Bulletin 103:446–457.

Conner, R. N., D. C. Rudolph, and J. R. Walters. 2001. The Red-cockaded Woodpecker: Surviving in a fire-maintained ecosystem. University of Texas Press, Austin. 363 pp.

Jackson, J. A. 1994. Red-cockaded Woodpecker (*Picoides borealis*). *In* The birds of North America, No. 85 (A. Poole and F. Gill, eds.). Academy of Natural Sciences, Philadelphia, PA, and American Ornithologists' Union, Washington, DC.

Mengel, R. M., and J. A. Jackson. 1977. Geographic variation in the Red-cockaded Woodpecker. Condor 79:349–355.

Shaffer, M. L. 1981. Minimum population sizes for species conservation. Bioscience 31:131–134.

———. 1987. Minimum viable populations: Coping with uncertainty. Pp. 69–86 *in* Viable populations for conservation (M. E. Soule, ed.). Cambridge University Press, Cambridge, UK.

US Fish and Wildlife Service. 2003. Recovery plan for the Red-cockaded Woodpecker (*Picoides borealis*): Second revision. US Fish and Wildlife Service, Atlanta, GA. 296 pp.

RED-CROWNED PARROT

Brush, T. 2005. Nesting birds of a tropical frontier: The Lower Rio Grande Valley of Texas. Texas A&M University Press, College Station. 262 pp.

Clinton-Eitniear, J., 1986. Status of the Green-cheeked Amazon in northeastern Mexico. Watchbird 13:22–34.

Enkerlin-Hoeflich, E. C. 1995. Comparative ecology and reproductive biology of three species of *Amazona* parrots in northeastern Mexico. PhD diss., Texas A&M University, College Station.

Gehlbach, F. R. 1987. Natural history sketches, densities, and biomass of breeding birds in evergreen forests of the Rio Grande, Texas and the Rio Corona, Tamaulipas, Mexico. Texas Journal of Science 39:241–251.

Iñigo-Elias, E. E., and M. A. Ramos. 1991. The psittacine trade in Mexico. Pp. 380–392 *in* Neotropical wildlife use and conservation (J. G. Robinson and K. H. Redford, eds.). University of Chicago Press, Chicago, IL. 538 pp.

Neck, R. W. 1986. Expansion of Red-crowned Parrots

(*Amazona viridigenalis*) into southern Texas and changes in agricultural practices in northern Mexico. Bulletin of the Texas Ornithological Society 19:6–12.

Patterson, S., and L. V. Lof. 2010. El Valle: The Rio Grande Delta. Gorgas Science Foundation, Brownsville, TX. 284 pp.

Chapter Ten

Buster, N. A., and C. W. Holmes. 2011. Gulf of Mexico origin, waters, and biota. Vol. 3, Geology. Harte Research Institute for Gulf of Mexico Studies Series. Texas A&M University Press, College Station. 446 pp.

Felder, D. L., and D. K. Camp. Gulf of Mexico origin, waters, and biota. Vol. 1, Biodiversity. Harte Research Institute for Gulf of Mexico Studies Series. Texas A&M University Press, College Station. 1393 pp.

CARIBBEAN ELECTRIC RAY

Bigelow, H. B., and W. C. Schroeder. 1953. Fishes of the western North Atlantic. Part 2: Sawfishes, guitarfishes, skates and rays, chimaeroids. Yale University Press, New Haven, CT. 588 pp.

Carrier, J. C., J. A. Musick, and M. R. Heithaus, eds. 2004. Biology of sharks and their relatives. CRC Press, Boca Raton, FL. 596 pp.

———, eds. 2012. Biology of sharks and their relatives. 2nd ed. CRC Press, Boca Raton, FL. 666 pp.

Dulvy, N. K., S. L. Fowler, J. A. Musick, et al. 2014. Extinction risk and conservation of the world's sharks and rays. eLife 3:e00590.

Fowler, S. L., R. D. Cavanagh, M. Camhi, et al., eds. 2005. Sharks, rays and chimaeras: The status of the chondrichthyan fishes. IUCN SSC Shark Specialist Group. IUCN, Gland, Switzerland, and Cambridge, UK. 462 pp.

Shepherd, T. D., and R. A. Myers. 2005. Direct and indirect fishery effects on small coastal elasmobranchs in the northern Gulf of Mexico. Ecology Letters 8:1095–1104.

SAWFISHES

Bigelow, H. B., and W. C. Schroeder, eds. 1953. Fishes of the western North Atlantic. Part 2: Sawfishes, guitarfishes, skates and rays, chimaeroids. Yale University Press, New Haven, CT. 588 pp.

Carlson, J. K., and C. A. Simpfendorfer. 2015. Recovery potential of Smalltooth Sawfish, *Pristis pectinata*, in the United States determined using population viability models. Aquatic Conservation: Marine and Freshwater Ecosystems 25:187–200.

Carrier, J. C., J. A. Musick, and M. R. Heithaus, eds. 2012.

Biology of sharks and their relatives. 2nd ed. CRC Press, Boca Raton, FL. 666 pp.

Dulvy, N. K., L. N. K. Davidson, P. M. Kyne, et al. 2016. Ghosts of the coast: Global extinction risk and conservation of sawfishes. Aquatic Conservation: Marine and Freshwater Ecosystems 26:134–153.

Dulvy, N. K., S. L. Fowler, J. A. Musick, et al. 2014. Extinction risk and conservation of the world's sharks and rays. eLife 3:e00590.

Faria, V. V., M. T. McDavitt, P. Charvet, et al. 2013. Species delineation and global population structure of critically endangered sawfishes (Pristidae). Zoological Journal of the Linnean Society 167:136–164.

Fowler, S. L., R. D. Cavanagh, M. Camhi, et al., eds. 2005. Sharks, rays and chimaeras: The status of the chondrichthyan fishes. IUCN SSC Shark Specialist Group. IUCN, Gland, Switzerland, and Cambridge, UK. 462 pp.

National Marine Fisheries Service. 2009. Recovery plan for Smalltooth Sawfish (*Pristis pectinata*). National Marine Fisheries Service. Silver Spring, MD. 102 pp.

HAMMERHEAD SHARKS

Carrier, J. C., J. A. Musick, and M. R. Heithaus, eds. 2010. Sharks and their relatives II: Biodiversity, adaptive physiology, and conservation. CRC Press, Boca Raton, FL. 746 pp.

———, eds. 2012. Biology of sharks and their relatives. 2nd ed. CRC Press, Boca Raton, FL. 666 pp.

Compagno, L. J. V. 1984. FAO species catalogue. Vol. 4, Sharks of the world: An annotated and illustrated catalogue of shark species to date. Part 2, Carcharhiniformes. FAO Fisheries Synopsis 4:251–655.

Dulvy, N. K., S. L. Fowler, J. A. Musick, et al. 2014. Extinction risk and conservation of the world's sharks and rays. eLife 3:e00590.

Ebert, D. A., S. Fowler, and L. Compagno, eds. 2013. Sharks of the world. Wild Nature Press, Plymouth, UK. 528 pp.

Fowler, S. L., R. D. Cavanagh, M. Camhi, et al., eds. 2005. Sharks, rays and chimaeras: The status of the chondrichthyan fishes. IUCN SSC Shark Specialist Group. IUCN, Gland, Switzerland, and Cambridge, UK. 462 pp.

Klimley, A. P. 1987. The determinants of sexual segregation in the Scalloped Hammerhead Shark, *Sphyrna lewini*. Environmental Biology of Fishes 18:27–40.

Miller, M. H., J. Carlson, P. Cooper, et al. 2013. Status review report: Scalloped Hammerhead Shark (*Sphyrna lewini*). US Department of Commerce, NOAA/NMFS, Silver Spring, MD. 125 pp.

National Marine Fisheries Service. 1993. Fishery management plan for sharks of the Atlantic Ocean. US Department of Commerce, NOAA/NMFS, Silver Spring, MD. 167 pp.

SEA TURTLES

Ernst, C. H., and J. E. Lovich. 2009. Turtles of the United States and Canada. 2nd ed. Johns Hopkins University Press, Baltimore, MD. 827 pp.

Lewison, B., L. Crowder, and D. J. Shaver. 2003. The impact of Turtle Excluder Devices and fisheries closures on Loggerhead and Kemp's Ridley strandings in the western Gulf of Mexico. Conservation Biology 17: 1089–1097.

Lutz, P. L., and J. A. Musick, eds. 1997. The biology of sea turtles. CRC Press, Boca Raton, FL. 432 pp.

Lutz, P. L., J. A. Musick, and J. Wyneken, eds. 2003. The biology of sea turtles. Vol. 2. CRC Press, Boca Raton, FL. 455 pp.

Metz, T. L., and A. M. Landry Jr. 2013. An assessment of Green Sea Turtle (*Chelonia mydas*) stocks along the Texas coast, with emphasis on the Lower Laguna Madre. Chelonian Conservation and Biology 12: 293–302.

Plotkin, P. T. 2007. Biology and conservation of Ridley Sea Turtles. Johns Hopkins University Press, Baltimore, MD. 356 pp.

Shaver, D. J. 1998. Sea turtle strandings along the Texas coast, 1980–94: Characteristics and causes of Texas marine strandings. NOAA Technical Report NMFS 143: 57–72.

Shaver, D. J., and C. W. Caillouet Jr. 2015. Reintroduction of Kemp's Ridley (*Lepidochelys kempii*) Sea Turtle to PINS, Texas and its connection to head-starting. Herpetological Conservation and Biology 10 (Symposium): 378–435.

Shaver, D. J., K. M. Hart, I. Fujisaki, et al. 2013. Foraging area fidelity for Kemp's Ridleys in the Gulf of Mexico. Ecology and Evolution 3:2002–2012. doi:10.1002/ece3.594.

Shaver, D. J., K. M. Hart, I. Fujisaki, et al. 2016. Migratory corridors of adult female Kemp's Ridley turtles in the Gulf of Mexico. Biological Conservation 196:158–167.

Wyneken, J., K. J. Lohmann, and J. A. Musick, eds. 2013. The biology of sea turtles. Vol. 3. CRC Press, Boca Raton, FL. 457 pp.

BALEEN WHALES

Aguilar, A., and C. H. Lockyer. 1987. Growth, physical maturity and mortality of Fin Whales (*Balaenoptera physalus*) inhabiting the temperate waters of the northeast Atlantic. Canadian Journal of Zoology 65: 253–264.

Clark, C. W. 1995. Application of US Navy underwater hydrophone arrays for scientific research on whales. Report of the International Whaling Commission 45: 210B212.

Hohn, A. A. 2009. Age estimation. Pp. 11–17 *in* Encyclopedia of marine mammals, 2nd ed. (W. F. Perrin, B. Würsig, and H. Thewissen, eds.). Academic Press, San Diego, CA. 1352 pp.

Jefferson, T. A., M. A. Webber, and R. L. Pitman. 2015. Marine mammals of the World: A comprehensive guide to their identification. 2nd ed. Elsevier, Amsterdam, the Netherlands. 616 pp.

Jensen, A. S., and G. K. Silber. 2003. Large whale ship strike database. US Department of Commerce, NOAA Technical Memorandum NMFS-OPR-25:1–37.

Würsig, B., T. A. Jefferson, and D. J. Schmidly. 2000. Marine mammals of the Gulf of Mexico. Texas A&M University Press, College Station. 256 pp.

SPERM WHALE

Jefferson, T. A., M. A. Webber, and R. L. Pitman. 2015. Marine mammals of the world: A comprehensive guide to their identification. 2nd ed. Elsevier, Amsterdam, the Netherlands. 616 pp.

Jochens, A., D. Biggs, K. Benoit-Bird, et al. 2008. Sperm Whale seismic study in the Gulf of Mexico: Synthesis report. US Department of the Interior, Minerals Management Service, New Orleans, LA. 341 pp.

Mullin, K. D., and G. L. Fulling. 2004. Abundance of cetaceans in the oceanic northern Gulf of Mexico, 1996–2001. Marine Mammal Science 20:787–807.

National Marine Fisheries Service. 2015. Sperm Whale (*Physeter macrocephalus*) 5-year review: Summary and evaluation. Silver Spring, MD. 58 pp.

Reeves, R. R., J. N. Lund, T. D. Smith, and E. A. Josephson. 2011. Insights from whaling logbooks on whales, dolphins, and whaling in the Gulf of Mexico. Gulf of Mexico Science 5:41–67.

Richter, C., J. Gordon, N. Jaquet, and B. Würsig. 2008. Social structure of Sperm Whales in the northern Gulf of Mexico. Gulf of Mexico Science 2:118–123.

Schulz, T. M., H. Whitehead, S. Gero, and L. Rendell. 2011. Individual vocal production in a Sperm Whale (*Physeter macrocephalus*) social unit. Marine Mammal Science 27:149–166.

Whitehead, H. 2003. Sperm Whales: Social evolution in the ocean. University of Chicago Press, Chicago, IL. 464 pp.

Würsig, B. 2017. Marine mammals of the Gulf of Mexico. Pp. 1489–1587 *in* Habitats and biota of the Gulf of Mexico: Before the Deepwater Horizon oil spill, vol. 2 (C. H. Ward, ed.). Springer Verlag GMBH, Heidelberg, Germany. 891 pp.

Würsig, B., S. K. Lynn, T. A. Jefferson, and K. D. Mullin. 1998. Behavior of cetaceans in the northern Gulf of Mexico relative to survey ships and aircraft. Aquatic Mammals 24:41–50.

WEST INDIAN MANATEE

Gunter, G. 1941. Occurrence of the manatee in the United States, with records from Texas. Journal of Mammalogy 22:60–64.

Husar, S. L. 1977. The West Indian Manatee (*Trichechus manatus*). US Fish and Wildlife Service Wildlife Resource Report 7:1–22.

Lefebvre, L. W., J. P. Reid, W. J. Kenworthy, and J. A. Powell. 2000. Characterizing manatee habitat use and seagrass grazing in Florida and Puerto Rico: Implications for conservation and management. Pacific Conservation Biology 5:289–298.

O'Shea, T. J., G. B. Rathbun, R. K. Bonde, et al. 1991. An epizootic of Florida Manatees associated with a dinoflagellate bloom. Marine Mammal Science 7:165–179.

Powell, J. A., and G. B. Rathbun. 1984. Distribution and abundance of manatees along the northern coast of the Gulf of Mexico. Northeast Gulf Science 7:1–28.

Reynolds, J. E., III. 1999. Efforts to conserve the manatees. Pp. 267–295 *in* Conservation and management of marine mammals (J. R. Twiss Jr. and R. R. Reeves, eds.). Smithsonian Institution Press, Washington, DC. 496 pp.

Chapter Eleven

Adam, P. J. 2004. *Monachus tropicalis*. Mammalian Species 747:1–9.

Baisre, J. A. 2013. Shifting baselines and the extinction of the Caribbean Monk Seal. Conservation Biology 27: 0927–935.

Brune, G. 1981. Springs of Texas. Vol. 1. Texas A&M University Press, College Station. 584 pp.

Creel, G. C. 1964. *Hemigrapsus estellinensis*: A new grapsoid crab from North Texas. Southwestern Naturalist 8:236–241.

Greenberg, J. 2014. A feathered river across the sky: The Passenger Pigeon's flight to extinction. Bloomsbury, New York. 304 pp.

Gunter, G. 1947. Sight records of the West Indian Seal, *Monachus tropicalis* (Gray), from the Texas Coast. Journal of Mammalogy 28:289–290.

Oberholser, H. C. 1974. The bird life of Texas. Vol. 1. University of Texas Press, Austin. 530 pp.

Schmidly, D. J., and R. D. Bradley. 2016. The mammals of Texas. 7th ed. University of Texas Press, Austin. 694 pp.

Schorger, A. W. 1955. The Passenger Pigeon: Its natural history and extinction. University of Wisconsin Press, Madison. 452 pp. [Reprint editions are available.]

Tanner, J. T. 1942. The Ivory-billed Woodpecker. National Audubon Society, New York. 144 pp. [Dover Publications publishes a reprint edition.]

CONTRIBUTORS

Robert Allen
East Texas Suboffice
US Fish and Wildlife Service
Nacogdoches, TX 75965
The findings and conclusions are those of the author
and do not necessarily represent the views of the US
Fish and Wildlife Service.

Loren K. Ammerman
Department of Biology
Angelo State University
San Angelo, TX 76909

David J. Berg
Department of Biology
Miami University
Oxford, OH 45056

Karl S. Berg
Department of Biology
University of Texas Rio Grande Valley
Brownsville, TX 78520

Timothy Bonner
Department of Biology
Texas State University
San Marcos, TX 78666

Timothy Brush
Department of Biology
University of Texas Rio Grande Valley
Edinburg, TX 78539

Julia C. Buck
Marine Science Institute
University of California, Santa Barbara
Santa Barbara, CA 93106

Brian R. Chapman
Texas Research Institute for Environmental Studies
Sam Houston State University
Huntsville, TX 77341

David A. Cimprich
Environmental Division
Fort Hood Military Installation
Fort Hood, TX 76544
The content contributed by this author does not
necessarily reflect the position or policy of the federal
government and no official endorsement should be
inferred.

Christopher Comer
Department of Forestry
Stephen F. Austin State University
Nacogdoches, TX 75962

Jerry L. Cook
Department of Biological Sciences
Sam Houston State University
Huntsville, TX 77341

Cody F. Craig
Department of Biology
Texas State University
San Marcos, TX 78666

J. Curtis Creighton
Department of Biological Sciences
Purdue University Calumet
Hammond, IN 46323

Alice F. Echelle
Department of Integrative Biology
Oklahoma State University
Stillwater, OK 74078

Anthony A. Echelle
Department of Integrative Biology
Oklahoma State University
Stillwater, OK 74078

Robert Edwards
Department of Biology
University of Texas Rio Grande Valley
Edinburg, TX 78539

David F. Ford
Ecological Specialists, Inc.
O'Fallon, MO 63366

Neil B. Ford
Department of Biology
University of Texas at Tyler
Tyler, TX 75799

William B. Godwin
Museum of Natural History
Sam Houston State University
Huntsville, TX 77341

Nicky M. Hahn
Department of Biology
Texas State University
San Marcos, TX 78666

Chad W. Hargrave
  Department of Biological Sciences
  Sam Houston State University
  Huntsville, TX 77341
Benjamin T. Hutchins
  Texas Parks and Wildlife Department
  Austin, TX 78744
  The findings and conclusions are those of the author
  and do not necessarily represent the views of the Texas
  Parks and Wildlife Department.
Mary Jones
  Department of Biology
  Angelo State University
  San Angelo, TX 76909
John Karges
  The Nature Conservancy
  Texas Field Office
  San Antonio, TX 78215
Andrew C. Kasner
  School of Math and Science
  Wayland Baptist University
  Plainview, TX 79072
William I. Lutterschmidt
  Director, Texas Research Institute for Environmental
  Studies
  Sam Houston State University
  Huntsville, TX 77341
Jeremy D. Maikoetter
  Department of Biology
  Texas State University
  San Marcos, TX 78666
Heather Mathewson
  Department of Wildlife, Sustainability and Ecosystem
  Sciences
  Tarleton State University
  Tarleton, TX 76402
Philip Matich
  Texas Research Institute for Environmental Studies
  Sam Houston State University
  Huntsville, TX 77341
James M. Mueller
  US Fish and Wildlife Service
  24518 E. FM 1431
  Marble Falls, TX 78654
  The findings and conclusions are those of the author
  and do not necessarily represent the views of the US
  Fish and Wildlife Service.

Paige A. Najvar
  Austin Ecological Services Field Office
  US Fish and Wildlife Service
  Austin, TX 78758
  The findings and conclusions are those of the author
  and do not necessarily represent the views of the US
  Fish and Wildlife Service.
Josh B. Pierce
  Southern Forest Research Station
  USDA Forest Service
  Nacogdoches, TX 75965
D. Craig Rudolph
  Southern Forest Research Station
  USDA Forest Service
  Nacogdoches, TX 75965
David S. Ruppel
  Department of Biology
  Texas State University
  San Marcos, TX 78666
Clifford E. Shackelford
  Texas Parks and Wildlife Department
  506 Hayter Street
  Nacogdoches, TX 75961
  The findings and conclusions are those of the author
  and do not necessarily represent the views of the Texas
  Parks and Wildlife Department.
Donna J. Shaver
  Padre Island National Seashore
  National Park Service
  Corpus Christi, TX 78480
Mary K. Skorrupa
  Texas Coastal Ecological Services Office
  US Fish and Wildlife Service
  Corpus Christi, TX 78411
  The findings and conclusions are those of the author
  and do not necessarily represent the views of the US
  Fish and Wildlife Service.
Mary M. Streich
  Padre Island National Seashore
  National Park Service
  Corpus Christi, TX 78480
Ned E. Strenth
  Department of Biology
  Angelo State University
  San Angelo, TX 76909
Michael E. Tewes
  Caesar Kleberg Wildlife Research Institute
  Texas A&M University–Kingsville
  Kingsville, TX 78363

Kim Withers
    Department of Life Sciences and Center for Coastal
    Studies
    Texas A&M University–Corpus Christi
    Corpus Christi, TX 78412

Bernd Würsig
    Department of Marine Biology
    Texas A&M University at Galveston
    Galveston, TX 77553

INDEX

Page numbers in *italics* indicate illustrations.